# Math Triumphs

## Foundations for Geometry

McGraw Hill Glencoe

## Photo Credits

**iv v** CORBIS; **vi** PunchStock; **vii viii** Getty Images; **1** David Frazier/CORBIS; **2–3** Bryn Lennon/Getty Images; **8** John Howard/Getty Images; **14** CORBIS; **23** Doug Menuez/Getty Images; **24** Andersen Ross/Getty Images; **28** PunchStock; **33** CORBIS; **36–37** PunchStock; **41** Getty Images; **46** The McGraw-Hill Companies, Inc./Gerald Wofford, photographer; **47** Getty Images/SW Productions; **48** John A. Rizzo/Getty Images; **58** Ariel Skelley/Getty Images; **61 62** Ryan McVay/Getty Images; **69** Brand X Pictures/PunchStock; **70–71** Larry French/Getty Images; **75** Ryan McVay/Getty Images; **80** SuperStock; **82** PunchStock; **92** CORBIS; **97** Comstock Images/PictureQuest; **101** Ryan McVay/Getty Images; **103** JUPITERIMAGES/Comstock Premium/Alamy; **105** CORBIS; **106–107** Jeff Cadge/Getty Images; **114** Thomas Barwick/Getty Images; **118** D. Hurst/Alamy; **120** Manchan /Getty Images; **125** Getty Images; **126** PunchStock; **130** Barry Austin Photography/Riser/Getty Images; **132** Ryan McVay/Getty Images; **143** Getty Images; **146–147** Getty Images; **152 156** CORBIS; **157** The McGraw-Hill Companies, Inc./ Ken Cavanagh photographer; **158 166** CORBIS; **168** PunchStock; **171** Ryan McVay/Getty Images; **172** Comstock Images/Alamy; **175** Getty Images; **181** Hill Street Studios/Getty Images; **183** Glen Allison/Getty Images; **184–185** Justin Horrocks/Getty Images; **189** Rim Light/PhotoLink/Getty Images; **194** Getty Images; **195** Steve Mason/Getty Images; **201** C Squared Studios/Getty Images; **203** Getty Images/Somos.

*The McGraw-Hill Companies*

 **Glencoe**

Copyright © 2010 The McGraw-Hill Companies, Inc. All rights reserved. No part of this publication may be reproduced or distributed in any form or by any means, or stored in a database or retrieval system, without the prior written consent of The McGraw-Hill Companies, Inc., including, but not limited to, network storage or transmission, or broadcast for distance learning.

Send all inquiries to:
Glencoe/McGraw-Hill
8787 Orion Place
Columbus, OH 43240-4027

ISBN: 978-0-07-890859-0
MHID: 0-07-890859-0

Printed in the United States of America.

1 2 3 4 5 6 7 8 9 10  066  17 16 15 14 13 12 11 10 09 08

*Math Triumphs: Foundations for Geometry*
*Student Edition*

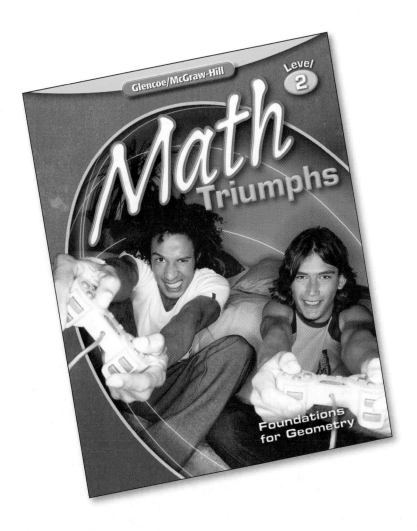

# Contents

## Chapter 1   Integers

*Albany, New York*

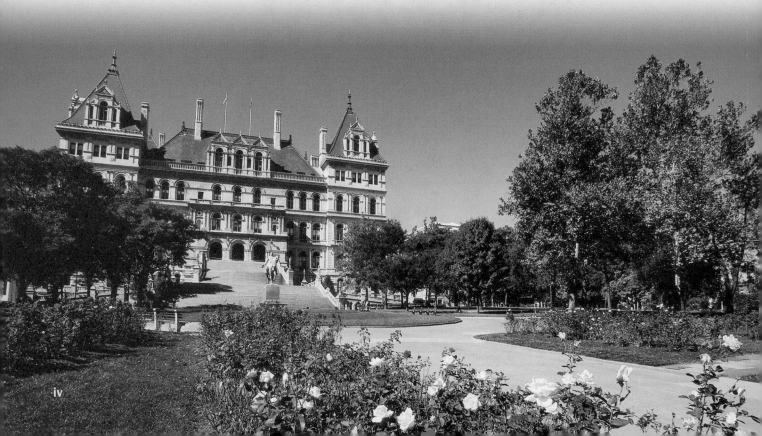

# Contents

## Chapter 2 — Real Numbers

*Yosemite National Park, California*

v

# Contents

## Chapter 3 — Equations and Inequalities

*Arches National Park, Utah*

# Chapter 4 — Linear Equations

*Caddo Lake, Texas*

# Contents

*Vietnam Veterans Memorial, Washington, D. C.*

**Chapter 6**

# Probability and Statistics

*Farm in Wisconsin*

## Heart Rate

You often reach your maximum heart rate when you exercise. You can evaluate the formula $220 - a$, where $a$ represents a person's age in years, to find your maximum heart rate.

STEP **2** **Preview**    Get ready for Chapter 1. Review these skills and compare them with what you will learn in this chapter.

| What You Know | What You Will Learn |
| --- | --- |
| You know how to add whole numbers.<br><br>**Example:** $3 + 8 = 11$<br><br>**TRY IT!**<br><br>**1**   $6 + 4 =$ ____    **2**   $13 + 7 =$ ____<br><br>**3**   $7 + 5 =$ ____    **4**   $12 + 12 =$ ____ | *Lesson 1-2*<br><br>You can use algebra tiles to add integers.<br><br>$$-4 + (-4) = -8$$<br>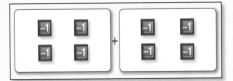 |
| You know how to multiply whole numbers.<br><br>**Example:** $6 \cdot 7 = 42$<br><br>**TRY IT!**<br><br>**5**   $6 \cdot 6 =$ ____    **6**   $8 \cdot 4 =$ ____<br><br>**7**   $10 \cdot 4 =$ ____    **8**   $7 \cdot 5 =$ ____ | *Lesson 1-3*<br><br>The product of two integers with the same signs is positive.<br><br>The product of two integers with different signs is negative.<br><br>$-13 \times 7 = -91$<br><br>The integers have different signs. The product is negative. |
| You know how to divide whole numbers.<br><br>**Example:** $100 \div 10 = 10$<br><br>**TRY IT!**<br><br>**9**   $63 \div 9 =$ ____   **10**   $72 \div 12 =$ ____<br><br>**11**   $28 \div 7 =$ ____   **12**   $20 \div 4 =$ ____ | *Lesson 1-3*<br><br>The quotient of two integers with the same sign is positive.<br><br>The quotient of two integers with different signs is negative.<br><br>$-21 \div (-3) = 7$<br><br>The integers have the same signs so the quotient is positive. |

# Number Properties

## KEY Concept

### Commutative Properties

**Addition**

$$3 + 4 = 4 + 3$$
$$7 = 7$$

The order of the addends changes, but the sum does not change.

$$a + b = b + a$$

**Multiplication**

$$2 \cdot 5 = 5 \cdot 2$$
$$10 = 10$$

The order of the factors changes, but the product does not change.

$$a \cdot b = b \cdot a$$

### Associative Properties

**Addition**

$$(5 + 4) + 3 = 5 + (4 + 3)$$
$$9 + 3 = 5 + 7$$
$$12 = 12$$

The order of the addends does not change, but the grouping changes.

$$(a + b) + c = a + (b + c)$$

**Multiplication**

$$(6 \cdot 5) \cdot 2 = 6 \cdot (5 \cdot 2)$$
$$30 \cdot 2 = 6 \cdot 10$$
$$60 = 60$$

The order of the factors does not change, but the grouping changes.

$$(a \cdot b) \cdot c = a \cdot (b \cdot c)$$

### Identity Properties

**Addition**

$$5 + 0 = 5$$
$$5 = 5$$

Any number plus 0 equals that number.

$$a + 0 = a$$

**Multiplication**

$$3 \cdot 1 = 3$$
$$3 = 3$$

Any number times 1 equals that number.

$$a \cdot 1 = a$$

### Multiplicative Property of Zero

$$9 \cdot 0 = 0$$
$$0 = 0$$

Any number times 0 equals 0.

$$a \cdot 0 = 0$$

### Distributive Property

$$4(1 + 3) = (4 \cdot 1) + (4 \cdot 3)$$
$$4(4) = 4 + 12$$
$$16 = 16$$

Multiply the number outside the parentheses by each number inside the parentheses.

$$a(b + c) = (a \cdot b) + (a \cdot c)$$

## VOCABULARY

**addend**
numbers or quantities being added together

**Associative Property of Addition**
the grouping of the addends does not change the sum

**Associative Property of Multiplication**
the grouping of the factors does not change the product

**Commutative Property of Addition**
the order in which two numbers are added does not change the sum

**Commutative Property of Multiplication**
the order in which two numbers are multiplied does not change the product

**Distributive Property**
to multiply a sum by a number, multiply each addend by the number outside the parentheses and add the products

**factor**
a number that divides into a whole number evenly

Number properties can be used to verify that expressions are equal.

## Example 1

**Verify the expressions are equal. Name the property shown.**

$$(7 \cdot 10) \cdot 4 = 7 \cdot (10 \cdot 4)$$

1. Simplify the expression on each side of the equation.

$$(7 \cdot 10) \cdot 4 = 7 \cdot (10 \cdot 4)$$
$$70 \cdot 4 = 7 \cdot 40$$
$$280 = 280$$

2. The factors are grouped differently, but the products are the same. The Associative Property of Multiplication is illustrated.

**YOUR TURN!**

**Verify the expressions are equal. Name the property shown.**

$$8 + 14 = 14 + 8$$

1. Simplify the expression on each side of the equation.

$$8 + 14 = 14 + 8$$
$$\underline{\hphantom{00}} = \underline{\hphantom{00}}$$

2. The _____ are in a different order, but the sums are _____. The

_____

is illustrated.

## Example 2

**Write an equation to show the Additive Identity Property.**

1. For addition, 0 is the identity.

2. Select a number to add to 0.
   23 + 0

3. Write an equation.
   23 + 0 = 23

**YOUR TURN!**

**Write an equation to show the Multiplicative Identity Property.**

1. For multiplication, _____ is the identity.

2. Select a number to multiply by _____.
   _____ $\cdot$ _____

3. Write an equation.
   _____ $\cdot$ _____ = _____

## Example 3

**Apply the Distributive Property to simplify 4(15 + 3).**

1. Multiply each addend by 4.
   $$4(15 + 3) = (4 \cdot 15) + (4 \cdot 3)$$
   $$= 60 + 12$$

2. Find the sum of the products.
   60 + 12 = 72

**YOUR TURN!**

**Apply the Distributive Property to simplify 8(12 + 9).**

1. Multiply each addend by _____.
   $$8(12 + 9) = (\underline{\hphantom{0}} \cdot \underline{\hphantom{0}}) + (\underline{\hphantom{0}} \cdot \underline{\hphantom{0}})$$
   $$= \underline{\hphantom{0}} + \underline{\hphantom{0}}$$

2. Find the sum of the products.
   _____ + _____ = _____

GO ON

**Guided Practice**

**Name each property shown.**

**1**  $42 \cdot 1 = 42$

_____

**2**  $11 + 9 = 9 + 11$

_____

**Step** (by) **Step Practice**

**3**  Verify the expressions are equal. Name the property shown.
$$15 + (3 + 25) = (15 + 3) + 25$$

**Step 1**  Simplify the expression on each side of the equation.

$15 + (3 + 25) =$ _____ + _____

$\phantom{15 + (3 + 25)} =$ _____

$(15 + 3) + 25 =$ _____ + _____

$\phantom{(15 + 3) + 25} =$ _____

> Notice the grouping of the addends is different on each side of the equation.

**Step 2**  The _____ are grouped differently, but the order has stayed _____.
The _____ is illustrated.

**Verify the expressions are equal. Name the property shown.**

**4**  $13 \cdot 2 \cdot 5 = 5 \cdot 2 \cdot 13$

_____

**5**  $8(4 + 10) = 112$

_____

**6**  $56 + 0 = 56$

_____

**7**  $1{,}071 \cdot 0 = 0$

_____

**Write an equation to show each property.**

**8**  Distributive Property

_____

**9**  Additive Identity

_____

**10**  Commutative Property of Multiplication

_____

**11**  Associative Property of Addition

_____

**Apply the Distributive Property to simplify each expression.**

**12**  $7(10 + 1)$

_____

**13**  $8(5 + 30)$

_____

**Solve.**

**14** TEST SCORES    Tenaya earned 82 points on one math test and 94 points on the next test. Elena earned 94 points on the first test and then 82 points on her next test. Compare the total test scores for Tenaya to the total test scores for Elena.

Tenaya: _____ + _____ = _____

Elena: _____ + _____ = _____

Check off each step.

_____ **Understand: I underlined the key words.**

_____ **Plan: To solve the problem, I will** _____

_____.

_____ **Solve: The answer is** _____.

_____ **Check: I checked my answer by** _____.

## Skills, Concepts, and Problem Solving

**Name each property shown.**

**15**   the total number of angles in 3 squares and 3 pentagons

$3(4 + 5) = 3 \cdot 4 + 3 \cdot 5$ _____

**16**   $(8 + 2) + 7 = 8 + (2 + 7)$ _____

**Verify the expressions are equal. Name the property shown.**

**17**   $0 \cdot 33 = 0$

_____

**18**   $3(21 \cdot 5) = (3 \cdot 21)5$

_____

**19**   $41 + 16 = 16 + 41$

_____

**20**   $2(13 + 9) = (2 \cdot 13) + (2 \cdot 9)$

_____

GO ON

**Write an equation to show each property.**

**21** Commutative Property of Addition

_____

**22** Distributive Property

_____

**23** Multiplicative Identity

_____

**24** Associate Property of Multiplication

_____

**Apply the Distributive Property to simplify each expression.**

**25** $12(4 + 6)$

_____

**26** $(7 + 7)14$

_____

**27** $5(12 + 9)$

_____

**28** $11(3 + 10)$

_____

**Solve.**

**29** RUNNING   Bruce runs five days a week. He runs four miles each time he runs. Jeff likes to run four days a week. He runs five miles each time he runs. Compare the total miles each of them run in a week.

_____

_____

**30** TYPING   Three assistants can type at the different speeds shown in the table. One day, Pam typed for 2 hours, Rick typed for 4 hours, and Taylor typed for 1 hour. How many total words did they type?

_____

| Name | Words per hour |
|--------|----------------|
| Pam | 3,900 |
| Rick | 4,800 |
| Taylor | 3,120 |

**Vocabulary Check**   **Write the vocabulary word that completes each sentence.**

**31** The _____ shows that adding zero to a number equals that number.

**32** The _____ shows that the grouping of the factors does not change the product.

**33** **Reflect**   Suppose you have homework in three classes: math, English and history. You can finish two before dinner and one after dinner. Does the order you complete the homework matter? Relate your answer to the number properties. _____

_____

STOP

# Add and Subtract Integers

## KEY Concept

When adding integers, there are two possibilities.

### Addends with the Same Signs

| 1. Find the absolute value of each addend. | $-7 + (-9)$ <br> $\lvert -7 \rvert = 7$ <br> $\lvert -9 \rvert = 9$ | |
|---|---|---|
| 2. Add the absolute values. | $7 + 9 = 16$ | |
| 3. The sum has the same sign as the original addends. | $-7 + (-9) = -16$ | Together, there are 16 negative tiles. |

### Addends with Different Signs

| 1. Find the absolute value of each addend. | $-5 + 13$ <br> $\lvert -5 \rvert = 5$ <br> $\lvert 13 \rvert = 13$ | |
|---|---|---|
| 2. Subtract the absolute values. | $13 - 5 = 8$ | |
| 3. The sum has the sign of the addend with the greatest absolute value. | $\lvert 13 \rvert > \lvert -5 \rvert$ <br> $-5 + 13 = 8$ | Remove 5 zero pairs to leave 8 positive tiles. |

When subtracting integers, change the sign of the addend following the subtraction sign to its opposite and change the operation sign to addition.

### VOCABULARY

**absolute value**
the distance a number is from zero on a number line

**addend**
numbers or quantities being added together

**difference**
the answer to a subtraction problem

**sum**
the answer to an addition problem

Subtracting an integer is the same as adding the opposite of that integer.

## Example 1

Find $15 + (-10)$.

1. Find the absolute value of each addend. $\qquad \lvert 15 \rvert = 15 \qquad \lvert -10 \rvert = 10$

2. Do the addends have the same signs? **no**
   Do you add or subtract the absolute values? **subtract**

3. Subtract the absolute values. $\qquad 15 - 10 = 5$

4. The sum will have the sign of the addend with the greatest absolute value.
   $\lvert 15 \rvert > \lvert -10 \rvert \qquad 15 + (-10) = 5$

**GO ON**

**YOUR TURN!**

**Find $-20 + (-9)$.**

1. Find the absolute value of each addend.     $|\underline{\qquad}| = \underline{\qquad}$     $|\underline{\qquad}| = \underline{\qquad}$

2. Do the addends have the same signs?     $\underline{\qquad\qquad}$
   Do you add or subtract the absolute values?     $\underline{\qquad\qquad}$

3. $\underline{\qquad\qquad}$ the absolute values.     $\underline{\qquad\qquad} = \underline{\qquad}$

4. The sum will have the $\underline{\qquad\qquad}$ as the addends.

   $-20 + (-9) = \underline{\qquad}$

## Example 2

**Solve the equation $z = -47 + 53 - 4$.**

1. Add the first two addends.
   The integers have different signs.

   $|47| = 47 \qquad |53| = 53$

   Subtract the absolute values.

   $53 - 47 = 6$

   The sum is positive because 53 has the greater absolute value.

   $-47 + 53 = 6$

2. Write the equation replacing $-47 + 53$ with 6. Then rewrite to add the opposite of 4.

   $z = 6 - 4$

   $z = 6 + (-4)$

3. The integers have different signs.

   $|6| = 6 \qquad |-4| = 4$

   Subtract the absolute values.

   $6 - 4 = 2$

   The sum will be positive because 6 has the greater absolute value.

   $z = -47 + 53 - 4 = 2$

**YOUR TURN!**

**Solve the equation $c = 81 + 63 - 10$.**

1. Add the first two addends.
   The integers have $\underline{\qquad\qquad}$.

   $|81| = \underline{\qquad} \qquad |63| = \underline{\qquad}$

   $\underline{\qquad\qquad}$ the absolute values.

   $\underline{\qquad\qquad}$

   The sum is $\underline{\qquad}$ because the

   addends $\underline{\qquad\qquad}$.

   $81 + 63 = \underline{\qquad}$

2. Write the equation replacing $81 + 63$ with 149. Then rewrite to add the opposite of 10.

   $c = 144 - 10$

   $c = 144 + (-10)$

3. The integers have $\underline{\qquad\qquad}$.

   $|144| = \underline{\qquad} \qquad |-10| = \underline{\qquad}$

   $\underline{\qquad\qquad}$ the absolute values.

   $144 - 10 = \underline{\qquad}$

   The sum will be $\underline{\qquad}$ because

   $\underline{\qquad}$ has the greater $\underline{\qquad\qquad}$.

   $c = 81 + 63 - 10 = \underline{\qquad}$

**Guided Practice**

**Find each sum or difference.**

**1** $-7 + 15 =$ _____

$|-7| =$ _____ and $|15| =$ _____

What will be the sign of the answer?

_____

**2** $14 - (-3) =$ _____

$|14| =$ _____ and $|-3| =$ _____

What will be the sign of the answer?

_____

**3** $-3 - 8 =$ _____

**4** $5 + (-11) =$ _____

**Solve each equation.**

**5** $z = 28 - 41$

The integers have _____ signs.

Do you add or subtract the absolute values? _____

The sign of the answer will be _____.

$|28| =$ _____ and $|-41| =$ _____

$41$ _____ $28 =$ _____

$z =$ _____

**6** $m = -39 + 66$

The integers have _____ signs.

Do you add or subtract the absolute values? _____

The sign of the answer will be _____.

$|-39| =$ _____ and $|66| =$ _____

$66$ _____ $39 =$ _____

$m =$ _____

**Step by Step Practice**

**7** Solve the equation $p = -23 + (-18) + 7$.

**Step 1** Add the first two addends.

The integers have _____.

_____ the absolute values.

The sum is _____ because the addends _____.

$|-23| =$ _____ $|-18| =$ _____

_____

$-23 + (-18) =$ _____

**Step 2** Write the equation replacing $-23 + (-18)$ with the sum.

$p =$ _____ $+ 7$

**Step 3** The integers have _____.

_____ the absolute values.

The sum will be _____ because _____ has the greater _____.

$|$_____$| =$ _____ $|7| = 7$

_____

$p = -23 + (-18) + 7$

$=$ _____

GO ON

**Solve each equation.**

**8** $b = -156 + 71$

The integers have _____, so _____ the absolute values.

$|-156| =$ _____ and $|71| =$ _____

$156$ _____ $71 =$ _____

Keep the sign of the integer with the _____.

$b =$ _____

**9** $c = 94 - 20 + 50$

$|$_____$| =$ _____ and $|$_____$| =$ _____ So, $94$ _____ $20 =$ _____

$c =$ _____ $+$ _____

$c =$ _____

**10** $v = 24 - 17$ _____ **11** $s = -18 - 22$ _____ **12** $r = -15 + 36 - 8$ _____

**13** $q = 11 - 28 - 19$ _____ **14** $p = 16 - (-36) + 1$ _____ **15** $x = -1 - (-76) + 50$ _____

## Step by Step Problem-Solving Practice

**Solve.**

**16** **ALLOWANCE** On Saturday, Andrea was paid a $10 allowance. She spent $7 for lunch. Later that day, she babysat and earned $18. How much money did Andrea have at the end of the day if she had no money on Saturday morning?

$A = 10 - 7 + 18$

$|$_____$| =$ _____ and $|$_____$| =$ _____ So, $10 - 7 =$ _____.

$A =$ _____ $+ 18$

$A =$ _____

Check off each step.

_____ Understand: I underlined the words.

_____ Plan: To solve the problem, I will _____.

_____ Solve: The answer is _____.

_____ Check: I checked my answer by _____

_____.

1

# Skills, Concepts, and Problem Solving

**Find each sum or difference.**

**17** $-9 - 43 = $ _____

**18** $-21 + 5 = $ _____

**19** $33 - 12 - 67 = $ _____

**20** $114 - 71 - (-9) = $ _____

**Solve each equation.**

**21** $n = -17 - 5$ _____

**22** $v = -51 + 16$ _____

**23** $h = -81 - 13 + 119$ _____

**24** $k = -64 + 5 - 32$ _____

**25** $a = -24 - 15 - 9$ _____

**26** $x = -234 + 186 - 224$ _____

**Solve.**

**27** BANKING   Larisa receives the bank statement below in the mail. If Larisa started with $25 in her account, how much money does she have in her account now? _____

| Date | Transaction Type | Amount |
|---|---|---|
| April 14 | Withdrawal | $15.00 |
| April 18 | Deposit | $55.00 |
| April 20 | Withdrawal | $35.00 |

**28** FOOTBALL   During a football game, Tony's team ran four plays in a row. They gained 12 yards, lost 5 yards, gained 1 yard and then lost 4 yards. What was the total yards gained or lost on those four plays? _____

**Vocabulary Check**   **Write the vocabulary word that completes each sentence.**

**29** The _____ of a number is the distance between the number and zero on the number line.

**30** The answer to a subtraction problem is called a _____.

**31** Reflect   Can the absolute value of a number be negative? Why or why not? Explain your answer.

_____

_____

_____

# Progress Check 1 (Lessons 1-1 and 1-2)

**Verify the expressions are equal. Name the property shown.**

**1** $4 \cdot 15 \cdot 5 = 4 \cdot 5 \cdot 15$

_____

_____

**2** $3(14 + 10) = (3 \cdot 14) + (3 \cdot 10)$

_____

_____

**Write an equation to show each property.**

**3** Associative Property of Addition

_____

**4** Multiplicative Identity

_____

**5** Commutative Property of Addition

_____

**6** Multiplication Property of Zero

_____

**Find each sum or difference.**

**7** $-23 - 8 =$ _____

**8** $-1 + 65 =$ _____

**9** $38 - 17 + 9 =$ _____

**10** $236 - 77 - (-4) =$ _____

**Solve each equation.**

**11** $g = -33 - 6$ _____

**12** $d = 89 - (-14)$ _____

**13** $y = 11 - 13 - 34$ _____

**14** $q = 163 + 28 - 88$ _____

**15** $m = -61 - 45 + 213$ _____

**16** $x = -10 + 75 - 22$ _____

**Solve.**

**17** **MAIL** Tina received three pieces of mail. One was a check for $134. Another was a bill for $46 and the last one was a bill for $89. What was the net balance of Tina's mail for the day?

_____

**18** **DIETING** Delford follows a diet plan. On Mondays, Wednesdays, and Fridays he eats a breakfast of 375 calories and a lunch of 455 calories. On Tuesdays, Thursdays, Saturdays, and Sundays he eats a breakfast of 455 calories and a lunch of 375 calories. What property ensures Delford gets the same number of calories each day?

_____

# Multiply and Divide Integers

## KEY Concept

- When two integers with the same signs are multiplied or divided, their **product** is positive.

- When two integers with different signs are multiplied or divided, their product is negative.

| Traditional Multiplication Method | Partial Products Method |
|---|---|
| 42<br>× 57<br>———<br>294<br>+ 2100<br>———<br>2,394 | 42<br>× 57<br>———<br>$7 \times 2 = 14$<br>$7 \times 40 = 280$<br>$50 \times 2 = 100$<br>$50 \times 40 = 2,000$<br>————<br>2,394 |

When dividing integers, use long division.

### VOCABULARY

**dividend**
the number that is being divided

**divisor**
the number by which the dividend is being divided

**product**
the answer to a multiplication problem

**quotient**
the answer or result of a division problem

Practice each multiplication method. Then choose the one that works best for you.

## Example 1

**Use partial products to multiply $17 \cdot (-15)$.**

1. Write the problem vertically.
$$\begin{array}{r} 17 \\ \times\, {-15} \\ \hline \end{array}$$

2. Multiply 5
   - by the ones column.    $5 \cdot 7 = 35$
   - by the tens column.    $5 \cdot 10 = 50$

3. Multiply 10
   - by the ones column.    $10 \cdot 7 = 70$
   - by the tens column.    $10 \cdot 10 = 100$

4. Add the partial products.

   $35 + 50 + 70 + 100 = 255$

5. The signs are different, so the product is negative.

   $17 \cdot (-15) = -255$

## YOUR TURN!

**Use partial products to multiply $-24 \cdot (-11)$.**

1. Write the problem vertically

2. Multiply _____
   - by the ones column. ____ • ____ = ____
   - by the tens column. ____ • ____ = ____

3. Multiply _____
   - by the ones column. ____ • ____ = ____
   - by the tens column. ____ • ____ = ____

4. Add the partial products.

   ____ + ____ + ____ + ____ = ____

5. The signs are the same, so the product is _____.

   $-24 \cdot (-11) =$ _____

GO ON

## Example 2

**Solve the equation $r = 235 \div (-8)$.**

1. Write the problem as long division.

2. Because the divisor is greater than the first digit of the dividend, look at the first two digits.

   $8\overline{)235}$

   8 goes into 23 two times.

   $$\begin{array}{r} 2\phantom{00} \\ 8\overline{)235} \\ -16\phantom{0} \\ \hline 75 \end{array}$$

3. Multiply and subtract. Then bring down the next digit in the dividend.

4. How many times does 8 divide into 75?

   8 goes into 75 nine times.

   $$\begin{array}{r} 29\,R3 \\ 8\overline{)235} \\ -16\phantom{0} \\ \hline 75 \\ -72 \\ \hline 3 \end{array}$$

5. Multiply and subtract. Write the remainder next to the quotient.

6. The integers have different signs, so the quotient is negative.

   $$r = 235 \div (-8) = -29\ R3$$

## YOUR TURN!

**Solve the equation $d = \dfrac{-814}{-9}$.**

1. Write the problem as long division.

2. Because the divisor is greater than the first digit of the dividend, look at the first two digits. _____

   _____.

   $$\begin{array}{r} \phantom{0}\overline{)\phantom{000}} \\ \underline{\phantom{000}} \\ \phantom{000} \\ \underline{\phantom{000}} \end{array}$$

3. Multiply and subtract. Then bring down the next digit in the dividend.

4. How many times does 9 divide into 4? _____

   _____.

5. Multiply and subtract. Write the remainder next to the quotient.

6. The integers have the

   _____, so the quotient

   is _____.

   $$d = \frac{-814}{-9} = \underline{\phantom{0000}}$$

---

## ▶ Guided Practice

**Find each product or quotient.**

**1**
$$\begin{array}{r} -12 \\ \times\ 25 \\ \hline \\ \end{array}$$

$$\begin{array}{r} + \underline{\phantom{000}} \\ \end{array}$$

The signs are _____, so

the answer is _____.

**2**  $198 \div (-5) =$ _____

The signs are _____, so

the answer is _____.

$$\begin{array}{r} 5\overline{)198} \\ -\underline{\phantom{000}} \\ \\ -\underline{\phantom{000}} \end{array}$$

## Step by Step Practice

**3** Solve the equation $y = -57 \cdot (-41)$.

**Step 1** Rewrite the problem vertically.

**Step 2** Multiply _____

   • by the ones column. _____

   • by the tens column. _____

**Step 3** Multiply _____

   • by the ones column. _____

   • by the tens column. _____

**Step 4** Add the partial products. _____ + _____ + _____ + _____ = _____

**Step 5** The signs are the _____, so the product is _____.

   $y = -57 \cdot (-41) =$ _____

**Solve each equation.**

**4** $z = 31 \cdot (-14)$ _____

**5** $f = \dfrac{542}{8}$ _____

## Step by Step Problem-Solving Practice

**Solve.**

**6** **WEATHER**   It was –3°F outside. The record low is 12 times as cold. What is the record low temperature?

Check off each step.

_____ Understand: I underlined key words.

_____ Plan: To solve the problem, I will _____.

_____ Solve: The answer is _____.

_____ Check: I checked my answer by _____.

GO ON

# ▶ Skills, Concepts, and Problem Solving

**Find each product or quotient.**

**7**  $16 \cdot (-12) = $ _____

**8**  $335 \div (-6) = $ _____

**9**  $189 \div 4 = $ _____

**10**  $-22 \cdot (-65) = $ _____

**Solve each equation.**

**11**  $w = 19 \cdot (-6)$  _____

**12**  $p = -24 \cdot (-12)$  _____

**13**  $s = 943 \div 7$  _____

**14**  $d = \dfrac{195}{13}$  _____

**15**  $g = 47 \cdot 11$  _____

**16**  $h = -851 \div (-9)$  _____

**Solve.**

**17**  **SCHOOL SUPPLIES**  Rachel buys 12 markers at the school store for $3.60. How much was each marker?

_____

**18**  **PAYDAY**  Yori earns $960 a week. How much money would Yori earn in a year?

_____

**19**  **AREA**  The area of a rectangle is the product of the length and the width of the rectangle. What is the area of this rectangle?

8 cm
53 cm

_____

**Vocabulary Check**  **Write the vocabulary word that completes each sentence.**

**20**  The answer to a multiplication problem is called the _____.

**21**  In a division problem, the _____ is the number that is being divided.

**22**  **Reflect**  What are the two ways to multiply two integers? Which method do you prefer? Explain your choice.

_____

_____

_____

**STOP**

# Variables and Expressions

## KEY Concept

When translating a verbal phrase into an **expression**, there is usually a 1-to-1 correspondence between the words and symbols.

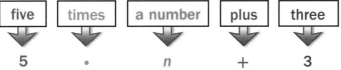

| five | times | a number | plus | three |
| 5 | · | $n$ | + | 3 |

The expression becomes $5n + 3$.

> The expression 5 · $n$ is written $5n$.

One exception to the 1-to-1 correspondence is the phrase "less than." You have to switch the order of the terms.

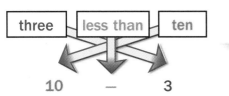

| three | less than | ten |
| 10 | − | 3 |

> "three less than ten" is 7, which is 10 take away three.

So, the expression should be written $10 − 3$.

### VOCABULARY

**expression**
a combination of numbers, variables, and operation symbols

**term**
a number, a variable, or a product or quotient of numbers and variables

**variable**
a symbol used to represent an unspecified number or value

Any letter or symbol can be used as a **variable**. The letters $i$, $l$, and $o$ are not good choices because they look too much like the numbers 1 and 0.

## Example 1

**Write an expression for the verbal phrase six more than twice a number.**

1. Group the phrase into parts.

| six | more than | two times | a number |

2. Translate.

   6     +     2 ·     $n$

3. Write the expression. $6 + 2n$

> $2 · n = 2n$

### YOUR TURN!

**Write an expression for the verbal phrase three less than half a number.**

> Remember *less than* is not a 1-to-1 translation. The order switches.

1. Group the phrase into parts.

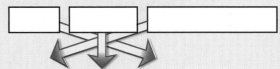

2. Translate.

_____ _____ _____

3. Write the expression. _____

**GO ON**

## Example 2

Neela bought three boxes of pens. Write an expression for the total number of pens she bought.

1. Write a verbal phrase.

   three times the number of pens per box

2. Group the phrase into parts.

   | three | times | number of pens per box |

   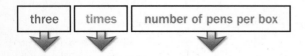

3. Translate.

   3      •               p

4. Write the expression. **3p**

   $3 \cdot p = 3p$

**YOUR TURN!**

Arleta received 8 more collectable stamps for her birthday. Write an expression for the total number of stamps in her collection.

1. Write a verbal phrase.

   _____.

2. Group the phrase into parts.

   |  |  |  |

   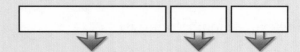

3. Translate.

   _____        _____ _____

4. Write the expression. _____

 ## Guided Practice

**Write an expression for each phrase.**

**1** eight more than a number

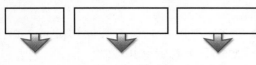

____   ____   ____

**2** twelve less than a number *b*.

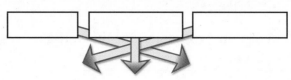

____ ____ ____

**3** the quotient of *a* and four

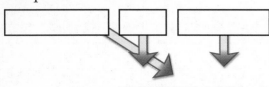

____ ____ ____

**4** the sum of eight and six times a number *y*

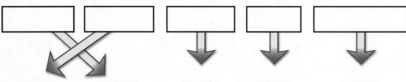

____   ____   ____   ____   ____

**5** Write an expression for the phrase.

eight less than three times a number

**Step 1** Group the phrase into parts.

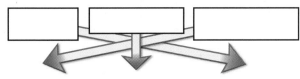

**Step 2** Translate.

_____    _____    _____

**Step 3** Write the expression. _____

**Write each number as a decimal.**

**6** the quotient of two times a number and 5 _____

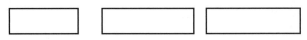

| the quotient of | two | times | a number | and 5 |

_____    _____    _____    _____    _____

**7** fifteen more than a number _____

_____    _____    _____

**8** the sum of 10 times $c$ and 7

_____

**9** Ricardo earns 3 dollars less than twice what Sara earns.

_____

**10** the product of $2k$ and 4

_____

**11** Geri received 4 more dollars than her sister.

_____

GO ON

**Solve.**

**12** **POSTCARDS** Simone bought 12 more than half the number of postcards in her collection. Write an expression to show how many postcards she has now.

The verbal phrase is 12 more than half the number of postcards.

_____   _____   _____   _____

Check off each step.

_____ **Understand: I underlined key words.**

_____ **Plan: To solve the problem, I will** _____.

_____ **Solve: The answer is** _____.

_____ **Check: I checked my answer by** _____

_____.

 **Skills, Concepts, and Problem Solving**

**Write an expression for each phrase.**

**13** 14 more than a number

_____

**14** 4 less than twice a number

_____

**15** the quotient of $p$ and ten

_____

**16** the sum of five times $b$ and 9

_____

**17** the difference of $m$ and 21

_____

**18** six less than 4 times $b$

_____

**19** José had 3 more hits today than in yesterday's game.

_____

**20** Roberto mows 10 more yards than twice the number Jim mows.

_____

**Solve.**

**21** SCHOOL PLAY   Tickets for the school play are $11 each. The money raised is the product of the number of tickets sold and the price of the ticket. Write an expression to represent the money raised from tickets sold for the school play.

_____

**22** AREA   The area of a rectangle is the product of its length and width. Write an expression for the area of the rectangle.

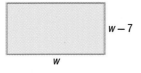

$w - 7$

$w$

_____

**23** CARS   Gabriella drove 250 miles on her last tank of gas. To calculate the miles per gallon from that tank of gas, she must find the quotient of 250 and the number of gallons she puts in her gas tank when she fills it up. Write an expression for the miles per gallon she got from her last tank of gas.

_____

**Vocabulary Check**   **Write the vocabulary word that completes each sentence.**

**24** $9x - 3$ is an example of an _____.

**25** A _____ is a letter or symbol used to represent an unspecified number.

**26** A(n) _____ is a number, a variable, or a product or quotient of numbers and variables.

**27** Reflect   Explain why both expressions, $10 + n$ and $n + 10$, are equivalent. Then tell how they represent the verbal phrase *10 more than number*.

_____

_____

_____

**Find each product or quotient.**

**1**   $102 \cdot (-9) =$ _____

**2**   $-136 \div (-5) =$ _____

**3**   $-47 \cdot 24 =$ _____

**4**   $-714 \div 6 =$ _____

**Solve each equation.**

**5**   $z = 9 \cdot (-56)$ _____

**6**   $y = -17 \cdot (-19)$ _____

**7**   $b = 881 \div 4$ _____

**8**   $n = \dfrac{-268}{12}$ _____

**9**   $t = 55 \cdot (-8)$ _____

**10**   $p = -315 \div (-15)$ _____

**Write an expression for each phrase.**

**11**   20 less than a number

_____

**12**   1 more than four times a number

_____

**13**   the product of a number and three

_____

**14**   the sum of twice $k$ and 3

_____

**Solve.**

**15**   **BOOK STORE**   Adam is buying books. He selected 8 books from a rack in the store. Then he realized the books were not marked with a price, but each book was the same price. Write an expression to represent the amount of money Adam needed to buy all 8 books.

_____

**16**   **ADVERTISING**   A billboard showed that a shopping mall was $x$ miles ahead. If the billboard was at mile marker 124, what expression represents the mile marker location of the shopping mall?

_____

# Order of Operations

## KEY Concept

When evaluating an expression with multiple operations, use the order of operations.

1. Simplify within parentheses or **grouping symbols**.

2. Simplify terms with exponents.

3. Multiply and divide from left to right.

4. Add and subtract from left to right.

| | |
|---|---|
| $-49 \div (5 + 2) \cdot 3^2 - 8$ | Simplify grouping symbols. |
| $= -49 \div 7 \cdot 3^2 - 8$ | Simplify exponents. |
| $= -49 \div 7 \cdot 9 - 8$ | Divide. |
| $= -7 \cdot 9 - 8$ | Multiply. |
| $= -63 - 8$ | Subtract. |
| $= -71$ | |

**VOCABULARY**

**grouping symbols**
symbols such as ( ), [ ], and { } that group two or more terms together

**operation**
mathematical process of calculation

---

## Example 1

**Evaluate $12 \div 4 + 5 \cdot 3 - 6$.**

1. Simplify grouping symbols.
   There are no grouping symbols.

2. Simplify exponents.
   There are no exponents.

3. Multiply and divide from left to right.

   $12 \div 4 + 5 \cdot 3 - 6$
   $= 3 + 5 \cdot 3 - 6$
   $= 3 + 15 - 6$

4. Add and subtract from left to right.

   $= 3 + 15 - 6$
   $= 18 - 6$
   $= 12$

   $12 \div 4 + 5 \cdot 3 - 6 = 12$

**YOUR TURN!**

**Evaluate $30 \div (5 - 2) \cdot 4$.**

1. Simplify grouping symbols.

   $30 \div (5 - 2) \cdot 4 = 30 \div \underline{\quad} \cdot 4$

2. Simplify exponents.

   There are \underline{\quad} exponents.

3. Multiply and divide from left to right.

   $30 \div 3 \cdot 4 = \underline{\quad} \cdot 4 = \underline{\quad}$

4. Add and subtract from left to right.

   There is \underline{\quad} addition or subtraction.

   $30 \div (5 - 2) \cdot 4 = \underline{\quad}$

## Example 2

**Evaluate $12 + (1 + 7^2) \div 5$.**

1. Simplify grouping symbols.

$12 + (1 + 7^2) \div 5$

> Follow the order of operations inside the grouping symbol.

$= 12 + (1 + 49) \div 5$

$= 12 + 50 \div 5$

2. Simplify exponents.
   There are **no** exponents outside of the grouping symbols.

3. Multiply and divide from left to right.

$= 12 + 50 \div 5$

$= 12 + 10$

4. Add and subtract from left to right.

$= 12 + 10 = 22$

$12 + (1 + 7^2) \div 5 = 22$

---

**YOUR TURN!**

**Evaluate $(2 \cdot 4)^2 - 3 \cdot (5 + 3)$.**

1. Simplify grouping symbols.

$(2 \cdot 4)^2 - 3 \cdot (5 + 3)$

$= (\underline{\quad})^2 - 3 \cdot \underline{\quad}$

2. Simplify exponents.

$= (8)^2 - 3 \cdot 8$

$= \underline{\quad} - 3 \cdot 8$

3. Multiply and divide from left to right.

$= 64 - 3 \cdot 8$

$= 64 - \underline{\quad}$

4. Add and subtract from left to right.

$= 64 - 24 = \underline{\quad}$

$(2 \cdot 4)^2 - 3 \cdot (5 + 3) = \underline{\quad}$

---

▶ ## Guided Practice

**Evaluate each expression.**

1. $25 \div (3 + 2) + 1 \cdot 6 = \underline{\quad}$
   Simplify grouping symbols.           $= 25 \div \underline{\quad} + 1 \cdot 6$
   Simplify exponents. There are \underline{\quad} exponents.
   Multiply and divide from left to right.   $= \underline{\quad} + \underline{\quad}$
   Add and subtract from left to right.    $= \underline{\quad}$

2. $8 \cdot (18 \div 3)^2 - 7 \cdot 6 = \underline{\quad}$
   Simplify grouping symbols.            $= 8 \cdot (\underline{\quad})^2 - 7 \cdot 6$
   Simplify exponents.                  $= 8 \cdot \underline{\quad} - 7 \cdot 6$
   Multiply and divide from left to right.   $= \underline{\quad} - \underline{\quad}$
   Add and subtract from left to right.    $= \underline{\quad}$

3. $(34 + 46) \div 20 + 20 = \underline{\quad}$
   Simplify grouping symbols.            $= \underline{\quad} \div 20 + 20$
   Simplify exponents. There are \underline{\quad} exponents.
   Multiply and divide from left to right.   $= \underline{\quad} + 20$
   Add and subtract from left to right.    $= \underline{\quad}$

## Step by Step **Practice**

**4** Evaluate $14 \div 7 \cdot (6 \div 2) + 10$.

| | | |
|---|---|---|
| **Step 1** | Simplify grouping symbols. | $14 \div 7 \cdot (6 \div 2) + 10$ |
| **Step 2** | Simplify exponents. There are ___ exponents. | $= 14 \div 7 \cdot (\underline{\phantom{xx}}) + 10$ |
| **Step 3** | Multiply and divide from ___ to ___. | $= \underline{\phantom{xx}} \cdot 3 + 10$ |
| **Step 4** | Add and subtract from ___ to ___. | $= \underline{\phantom{xx}} + 10$ |
| | | $= \underline{\phantom{xx}}$ |

**Evaluate each expression.**

**5** $7 - (8 + 3^2) + 1 = \underline{\phantom{xxx}}$

$\underline{\phantom{xx}} - (\underline{\phantom{xx}} + \underline{\phantom{xx}}) + \underline{\phantom{xx}}$

$\underline{\phantom{xx}} - (\underline{\phantom{xx}}) + \underline{\phantom{xx}}$

$\underline{\phantom{xx}} + \underline{\phantom{xx}} = \underline{\phantom{xx}}$

**6** $7 \cdot (6 \cdot 6 + 11) + 5 \cdot 4 = \underline{\phantom{xxx}}$

**7** $4 + (2^3 - 6 \cdot 3) + 2 \cdot 3 = \underline{\phantom{xxx}}$

## Step by Step **Problem-Solving Practice**

**Solve.**

**8** Barb is paid $15 an hour up to 40 hours a week. If Barb works more than 40 hours, she is paid $22 an hour for each hour over 40. Barb worked 46 hours this week. How much did Barb earn?

Number of hours that Barb worked over 40 hours: $46 - 40 = 6$

$(\underline{\phantom{xx}} \cdot 15) + (\underline{\phantom{xx}} \cdot 6) = \underline{\phantom{xx}} + \underline{\phantom{xx}} = \underline{\phantom{xx}}$

Check off each step.

_____ **Understand: I underlined key words.**

_____ **Plan: To solve the problem, I will** _____.

_____ **Solve: The answer is** _____.

_____ **Check: I checked my answer by** _____

_____.

**GO ON**

# ▶ Skills, Concepts, and Problem Solving

**Evaluate each expression.**

**9** $(2 - 8) - (3^2 \cdot 1) - 7 = $ _____

**10** $12 \cdot 2 \cdot (10 \div 2 + 20) - 100 = $ _____

**11** $(9^2 \div 3) \cdot 3 + 8 = $ _____

**12** $16 - 3 \cdot (8 - 3)^2 \div 5 = $ _____

**13** $14 + (8 \div 2^2) - 5 = $ _____

**14** $(50 - 8) + 42 + (27 \div 3) = $ _____

**Solve.**

**15** **AREA** The area of a trapezoid is equal to the expression $\frac{1}{2}h(b_1 + b_2)$, where $b_1$ and $b_2$ are the parallel bases of the trapezoid. What is the area of the trapezoid shown if the height is 4 ft?

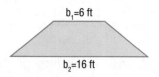

_____

**16** **LOGIC** Oscar's teacher gave him the following problem: $5 \cdot 4 \_\_ (3 \_\_ 2) - 1 = 3$. His assignment was to fill in the blanks with two operation signs that make the equation true. What should Oscar put in the blanks?

**$5 \cdot 4$ _____ $(3$ _____ $2) - 1 = 3$**

**17** **COOKING** Jeremy is cooking for his family tonight. He knows that he needs 1 cup more than double the amount of flour called for in the recipe. If the recipe called for 3 cups of flour, how many cups of flour does Jeremy need?

_____

**Vocabulary Check** **Write the vocabulary word that completes each sentence.**

**18** The _____ is a set of rules that tells what order to follow when evaluating an expression.

**19** Symbols such as ( ), { }, and [ ] are called _____.

**20** **Reflect** Write an expression using at least 3 different operations. Simplify it once using the order of operations. Then simplify it again going from left to right. Compare your answers. Explain why it is important to follow the order of operations.

_____

_____

_____

**STOP**

# Evaluate Expressions

## KEY Concept

**Expressions** can be evaluated for a given value of a **variable**.

$$2y - 4$$

Evaluate when $y = -2$.

$$2(-2) - 4 = -4 - 4 = -8$$

Evaluate when $y = 10$.

$$2(10) - 4 = 20 - 4 = 16$$

Evaluate when $y = 0.5$.

$$2(0.5) - 4 = 1 - 4 = -3$$

**VOCABULARY**

**expression**
a combination of numbers, variables, and operation symbols

**variable**
a letter or symbol used to represent an unknown quantity

When evaluating expressions with more than one variable, replace each variable with the given value. Then follow the order of operations to simplify. In geometry, formulas contain expressions that are evaluated to find a characteristic of a figure.

## Example 1

Evaluate the expression $d\left(a + \dfrac{b}{c}\right)$ when $a = 1$, $b = 4$, $c = 2$, and $d = 3$.

1. Rewrite the expression leaving spaces for the variables.

$$d\left(a + \frac{b}{c}\right) = (\quad)\left((\quad) + \frac{(\quad)}{(\quad)}\right)$$

2. Substitute the value of each variable into the corresponding space.

$$3\left(1 + \frac{4}{2}\right)$$

3. Follow the order of operations to simplify.

$$3\left(1 + \frac{4}{2}\right) = 3(1 + 2) = 3(3) = 9$$

### YOUR TURN!

Evaluate the expression $\dfrac{-3x + y}{z}$ when $x = 1$, $y = -2$, and $z = 10$.

1. Rewrite the expression leaving spaces for the variables.

$$\frac{-3x + y}{z} = \frac{-3\,\boxed{\phantom{x}} + \boxed{\phantom{x}}}{\boxed{\phantom{x}}}$$

2. Substitute the value of each variable into the corresponding space.

$$\frac{-3(\underline{\quad}) + (\underline{\quad})}{(\underline{\quad})}$$

3. Follow the order of operations to simplify.

$$\frac{\boxed{\phantom{x}} + \boxed{\phantom{x}}}{\boxed{\phantom{x}}} = \frac{\boxed{\phantom{x}}}{\boxed{\phantom{x}}} = \frac{\boxed{\phantom{x}}}{\boxed{\phantom{x}}}$$

GO ON

## Example 2

The formula for the area of a triangle is
$A = \frac{1}{2} bh$, where $b$ is the length of the
base and $h$ is the height of the triangle.

Find the area of a triangle when $b = 6$ in.
and $h = 3$ in.

1. Rewrite the formula leaving spaces for
   the variables.

   $A = \frac{1}{2} bh$

   $A = \frac{1}{2} (\ \ )(\ \ )$

2. Substitute the value of each variable into
   the corresponding space.

   $A = \frac{1}{2} (6)(3)$

3. Follow the order of operations to simplify.

   $A = \frac{1}{2} (6)(3) = 3(3) = 9$ in²

### YOUR TURN!

The formula for distance traveled is $d = rt$,
where $r$ is the rate and $t$ is the time.

Find the distance traveled at a rate of
55 mi/h for 3 hours.

1. Rewrite the formula leaving spaces for
   the variables.

   $d = rt$

   $d = \underline{\hspace{2cm}}$

2. Substitute the value of each variable into
   the corresponding space.

   $d = (\underline{\hspace{1cm}})(\underline{\hspace{1cm}})$

3. Follow the order of operations to
   simplify.

   $d = \underline{\hspace{3cm}}$

 **Guided Practice**

**Evaluate each expression.**

**1** $st \div 3$, when $s = 4$; $t = 6$

$(\underline{\hspace{1cm}})(\underline{\hspace{1cm}}) \div 3 = \underline{\hspace{1cm}}$

**2** $f(6 + g) + 1$, when $f = 2$; $g = 4$

$(\underline{\hspace{1cm}})(6 + (\underline{\hspace{1cm}})) + 1 = \underline{\hspace{1cm}}$

**3** $20 - \frac{2b}{c}$, when $b = 2$; $c = 4$

$20 - \dfrac{2(\boxed{\phantom{x}})}{(\boxed{\phantom{x}})} = \underline{\hspace{1cm}}$

**4** $a - b + 14$, when $a = 23$; $b = 5$ $\underline{\hspace{1cm}}$

$(\underline{\hspace{1cm}}) - (\underline{\hspace{1cm}}) + 14 = \underline{\hspace{1cm}}$

**5** The formula for the area of a rectangle is $A = \ell w$, where $\ell$ is the
length and $w$ is the width. Find the area of a rectangle when
$\ell = 15$ cm and $w = 20$ cm.

$A = \ell w$

$A = (\underline{\hspace{1cm}})(\underline{\hspace{1cm}})$

$A = \underline{\hspace{2cm}}$

**6** Evaluate the expression $ac + b\left(\dfrac{c}{d}\right)$ when $a = 4$, $b = 3$, $c = -10$, and $d = 2$.

**Step 1** Substitute the value of each variable into the expression.

$$ac + b\left(\frac{c}{d}\right) = (\underline{\quad})(\underline{\quad}) + (\underline{\quad})\left(\frac{(\boxed{\phantom{x}})}{(\boxed{\phantom{x}})}\right)$$

**Step 2** Follow the order of operations to simplify.

$$\underline{\hspace{4cm}} = \underline{\hspace{4cm}} = \underline{\hspace{4cm}}$$

**Evaluate each expression.**

**7** $3x - 2y$, when $x = 5$; $y = -7$

$3(\underline{\quad}) - 2(\underline{\quad})$

_____

**8** $2x \div y + 8$, when $x = 9$; $y = -3$

$2(\underline{\quad}) \div (\underline{\quad}) + 8$

_____

**9** $2(\ell + w)$, when $\ell = 6.5$; $w = 6$

_____

_____

**10** $5 - 3z^2$ when $z = 4$

_____

_____

**Solve.**

**11** **TEMPERATURE** The formula to convert degrees Celsius to degrees Fahrenheit is $F = \dfrac{9}{5}C + 32$.

Nolan measures the temperature of a mixture in science class as 30°C. What is the temperature of the mixture in degrees Fahrenheit?

$F = \dfrac{9}{5}C + 32 = \dfrac{9}{5}\underline{\hspace{1.5cm}} + 32 = \underline{\hspace{2cm}}$     The mixture is _____.

Check off each step.

_____ **Understand: I underlined key words.**

_____ **Plan: To solve the problem, I will** _____.

_____ **Solve: The answer is** _____.

_____ **Check: I checked my answer by** _____.

## Skills, Concepts, and Problem Solving

**Evaluate each expression when $x = 2$, $y = -3$, and $z = -4$.**

**12** $3x(y + z) = $ _____

**13** $-8y \div x + z = $ _____

**14** $x \cdot z \div 8 = $ _____

**15** $y + z \div x = $ _____

**16** $2y - 12 + 4x = $ _____

**17** $\frac{1}{2}xz + 7 = $ _____

**18** $15 - \frac{5x}{z} = $ _____

**19** $x \cdot y + \frac{3}{4}z = $ _____

**20** $5x - 3y + 2z = $ _____

**Solve.**

**21** **AREA** The length of a rectangular floor tile is 6 inches. The width is 8 inches. The formula for the area of a rectangle is $A = \ell w$, where $\ell$ is the length and $w$ is the width. What is the area of the tile?

_____

**22** **WEATHER** Samuel wants to convert 10°C to degrees Fahrenheit so that he can decide what type of coat to wear. The formula is $F = \frac{9}{5}C + 32$. What is the temperature in degrees Fahrenheit?

_____

**23** **PAINTING** Gage plans to paint a trapezoidal bedroom wall. He needs to find the area of the wall. The ceiling is 10 feet high. The formula for the area of a trapezoid is $A = \frac{1}{2}h(b_1 + b_2)$, where $h$ is the height, $b_1$ is the length of one base, and $b_2$ is the length of the other base. What is the area of his wall as shown at the right?

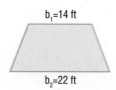

$b_1$=14 ft

$b_2$=22 ft

_____

**Vocabulary Check** **Write the vocabulary word that completes each sentence.**

**24** In the expression $2x + 5$, $x$ is called a _____.

**25** An _____ is a combination of numbers, variables, and operation symbols.

**26** **Reflect** Write all the possible integral side lengths for a rectangle that has an area of 18 cm². If a specific 18 cm² rectangle is desired, explain why you need to know more than the area to determine the side lengths.

_____

_____

**Evaluate each expression.**

**1** $(12 - 7) \div (2^2 + 1) - 9 =$ _____

**2** $6 \cdot 4 - (60 \div 5 \cdot 2) \cdot 100 =$ _____

**3** $(4^2 + 5) \div (9 - (3 - 1)) =$ _____

**4** $10 - 4 \cdot [(5 - 1)^2 \div 4] =$ _____

**5** $[2 - (7^2 - 7)] \div 5 =$ _____

**6** $10^2 + 5 \cdot 4 - 6 =$ _____

**7** $56 + (8 \div 2^3) - 5 =$ _____

**8** $(10 - 15) + 2 + (36 \div 3) =$ _____

**Evaluate each expression when $c = -2$, $d = -5$, and $f = 1$.**

**9** $2f(c + d) =$ _____

**10** $-d \div f + c =$ _____

**11** $2d \cdot c \div 4 =$ _____

**12** $5d + c \cdot 8f =$ _____

**13** $f - 10 + 3c =$ _____

**14** $\frac{1}{2}c + 8f =$ _____

**15** $-3 - \dfrac{10f}{c} =$ _____

**16** $d \cdot c + 12f \cdot c =$ _____

**17** $4c - d + f^2 =$ _____

**Solve.**

**18 AREA** The height of a triangle is 16 cm. The base of the triangle is 22 cm. The formula for the area of a triangle is $A = \frac{1}{2}bh$, where $h$ is the length and $b$ is the base. What is the area of the triangle?

_____

**19 HIKING** Jeremy is hiking for the third day in a row. On the first day, he hiked 6 miles. On the second day, he hiked 2 miles less than 2 times the distance he hiked the first day.

He wants to hike 3 miles more than half the sum of the miles of the first and second days. How many miles will he have hiked altogether at the end of the third day?

_____

# Chapter 1 Test

**Verify the expressions are equal. Name the property shown.**

**1** $(45 + 13) + 12 = 45 + (13 + 12)$

_____

**2** $72 \cdot 0 = 0$

_____

**Write an equation to show each property.**

**3** Distributive Property

_____

**4** Multiplicative Identity

_____

**5** Commutative Property of Addition

_____

**6** Associative Property of Multiplication

_____

**Apply the Distributive Property to simplify each expression.**

**7** $2(26 + 5)$

_____

**8** $7(2 + 11)$

_____

**Find each sum or difference.**

**9** $-15 - 78 =$ _____

**10** $-56 + 13 =$ _____

**11** $24 - 8 - (-101) =$ _____

**12** $511 - 97 + 7 =$ _____

**Find each product or quotient.**

**13** $-9 \cdot 55 =$ _____

**14** $-93 \cdot 4 =$ _____

**15** $303 \div (-5) =$ _____

**16** $-156 \div (-4) =$ _____

**Solve each equation.**

**17** $n = -12 \cdot (-8)$ _____

**18** $p = -42 \cdot 6$ _____

**19** $c = 516 \div 3$ _____

**20** $t = \dfrac{-89}{4}$ _____

**21** $w = -108 + 14$ _____

**22** $z = -51 \div 17$ _____

**23** $d = 107 - 79 + 4$ _____

**24** $h = -647 + 39 - 111$ _____

**Write an expression for each phrase.**

**25** 23 more than a number

_____

**26** the sum of two and a number divided by 3

_____

**27** product of a number and half that number

_____

**28** ten less than $b$

_____

**Evaluate each expression.**

**29** $(5 - 9) + (4^2 \cdot (-2)) =$ _____

**30** $2 \cdot 2 + (10 \cdot 20) - 1 =$ _____

**31** $(2^5 \cdot 16) \cdot 9 - 2 =$ _____

**32** $125 \div 5 \cdot 5 - 18 \div 3 =$ _____

**33** $8 + 8 - 8 \cdot 8 \div 8 =$ _____

**34** $1^2 - 2^3 \div 4^2 - 5^2 =$ _____

**35** $2(3a - b)$, when $a = -2$ and $b = 4$

_____

**36** $2 - c^2 + (-d)$ when $c = 1$ and $d = 6$

_____

**Solve.**

**37** **AREA**  The area of a trapezoid is equal to the expression $\frac{1}{2}h\,(b_1 + b_2)$, where $b_1$ and $b_2$ are the parallel bases of the trapezoid. What is the area of the trapezoid shown if the height is 5 inches?

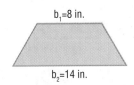

$b_1$=8 in.

$b_2$=14 in.

_____

**38** **BANKING**  Creag uses the Internet to check the daily transactions on his checking account. The report is shown at the right. If his balance at the beginning of the day was $212, what was his balance at the end of the day?

| Transaction Type | Amount |
|---|---|
| ATM Withdrawal | $45.00 |
| Deposit | $305.00 |
| Check 2345 payment | $124.00 |
| Check 2347 payment | $267.00 |

_____

**Correct the mistake.**

**39**  Regina simplified the expression $6 + 2^2 \times 14 - 10 \div 2$. Her result was 65. Did she use the order of operations correctly? If not, what is the correct answer?

STOP

_____

# Real Numbers

## Artists and architects often use the Golden Ratio.

The Golden Ratio is an irrational number that can be shown as 1.6108… Famous structures, such as the Parthenon in Greece, were designed using approximations of the Golden Ratio. The ratio of the width of the Parthenon to its height represents the use of the Golden Ratio in architecture.

STEP **2** **Preview**   Get ready for Chapter 2. Review these skills and compare them with what you will learn in this chapter.

| What You Know | What You Will Learn |
|---|---|
| You know how to multiply a number with two identical factors. | *Lesson 2-5* |

**What You Know**

You know how to multiply a number with two identical factors.

**Example:** $7 \cdot 7 = 49$

**TRY IT!**

1   $8 \cdot 8 =$ _____

2   $9 \cdot 9 =$ _____

3   $10 \cdot 10 =$ _____

4   $11 \cdot 11 =$ _____

**What You Will Learn**

*Lesson 2-5*

The **square root** of a number is one of the two equal factors of a number.

Find $\sqrt{4}$.

There are 2 tiles on each side of the square.

So, $\sqrt{4} = 2$.

---

You know how to graph whole numbers on a number line.

**Example:** Graph 1 and 4.

**TRY IT!**

5   Graph 3 and 5.

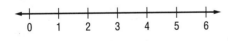

*Lesson 2-6*

To compare **real numbers**, use a number line.

Use $>$, $<$, or $=$ to compare 3.6 and $\sqrt{9}$.

So, $3.6 > \sqrt{9}$.

# Rational and Irrational Numbers

## KEY Concept

Real numbers include rational and irrational numbers.
A **rational number** can be written as a ratio of two integers.

| Real Numbers | |
|---|---|
| **Rational Numbers** | **Example** |
| fractions | $\dfrac{40}{100}$ |
| terminating decimals | $0.4 = \dfrac{40}{100}$ |
| repeating decimals | $0.\overline{3} = \dfrac{1}{3}$ |
| integers | $\sqrt{16} = 4 = \dfrac{4}{1}$ |

An **irrational number** cannot be written as a ratio of two integers. Irrational numbers are never-ending, never repeating decimals. For example, certain square roots and the value of pi are irrational numbers.

The symbol $\approx$ means that this is an estimation.

$$\pi \approx 3.141592654\ldots$$
$$\sqrt{23} \approx 4.79583\ldots$$

### VOCABULARY

**irrational number**
a number that cannot be written as a ratio of two integers

**rational number**
any number that can be written as a fraction $\dfrac{a}{b}$, where $a$ and $b$ are integers and $b \neq 0$

**real number**
a rational or an irrational number

Even though irrational numbers cannot be written as ratios, their values can be estimated.

## Example 1

**Is 0.25 rational or irrational?**

1. Is 0.25 a decimal, fraction, or radical?
   decimal

2. Can 0.25 be rewritten as a ratio with an integer in the numerator and a non-zero integer in the denominator?
   Yes; $0.25 = \dfrac{25}{100}$

3. 0.25 is a rational number.

## YOUR TURN!

**Is $\sqrt{12}$ rational or irrational?**

1. Is $\sqrt{12}$ a decimal, fraction, or radical?
   _____

2. Can $\sqrt{12}$ be written as a ratio with an integer in the numerator and a non-zero integer in the denominator?

   If a number is a perfect square, its square root is rational.

   _____

3. $\sqrt{12}$ is _____.

## Example 2

**Estimate the value of $\sqrt{58}$.**

1. Write an inequality using perfect square roots.

$$\sqrt{49} < \sqrt{58} < \sqrt{64}$$

2. Find the square root of the perfect squares.

$$\sqrt{49} = 7 \qquad \sqrt{64} = 8$$

3. The $\sqrt{58}$ is between 7 and 8.
   Is 58 closer to 49 or 64?

$$58 - 49 = 9 \qquad 64 - 58 = 6$$

   58 is closer to 64.

4. Because 58 is closer to 64, $\sqrt{58}$ is between 7 and 8, but closer to 8.

## YOUR TURN!

**Estimate the value of $\sqrt{30}$.**

1. Write an inequality using perfect square roots.

$$\underline{\hspace{1.5cm}} < \sqrt{30} < \underline{\hspace{1.5cm}}$$

2. Find the square root of the perfect squares.

$$\sqrt{25} = \underline{\hspace{1cm}} \qquad \sqrt{36} = \underline{\hspace{1cm}}$$

3. The $\sqrt{30}$ is between \_\_\_\_ and \_\_\_\_.
   Is 30 closer to \_\_\_\_ or \_\_\_\_?

$$30 - \underline{\hspace{0.8cm}} = \underline{\hspace{0.8cm}} \qquad 36 - \underline{\hspace{0.8cm}} = \underline{\hspace{0.8cm}}$$

   30 is closer to \_\_\_\_.

4. Because 30 is closer to \_\_\_\_, $\sqrt{30}$ is between 5 and 6, but closer to \_\_\_\_.

 **Guided Practice**

**Circle the word that classifies each number.**

**1** 0.125

Can it be written as a ratio? _____

rational        irrational

**2** $\sqrt{50}$

Can it be written as a ratio? _____

rational        irrational

**3** $\dfrac{4}{9}$

Can it be written as a ratio? _____

rational        irrational

**4** $\sqrt{81}$

Can it be written as a ratio? _____

rational        irrational

**5** Estimate the value of $\sqrt{95}$.

**Step 1** Write an inequality using perfect square roots.

$$\underline{\hspace{1.5cm}} < \sqrt{95} < \underline{\hspace{1.5cm}}$$

**Step 2** Find the square root of the perfect squares.

$$\sqrt{81} = \underline{\hspace{1.5cm}} \qquad \sqrt{100} = \underline{\hspace{1.5cm}}$$

**Step 3** The $\sqrt{95}$ is between $\underline{\hspace{1.5cm}}$ and $\underline{\hspace{1.5cm}}$.

Is 95 closer to $\underline{\hspace{1.5cm}}$ or $\underline{\hspace{1.5cm}}$?

$$95 - \underline{\hspace{1.5cm}} = \underline{\hspace{1.5cm}} \qquad \underline{\hspace{1.5cm}} - 95 = \underline{\hspace{1.5cm}}$$

95 is closer to $\underline{\hspace{1.5cm}}$.

**Step 4** Because 95 is closer to $\underline{\hspace{1.5cm}}$, $\sqrt{95}$ is between $\underline{\hspace{1.5cm}}$

and $\underline{\hspace{1.5cm}}$, but closer to $\underline{\hspace{1.5cm}}$.

**Estimate the value of each number.**

**6** $\sqrt{75}$

$$\underline{\hspace{1.5cm}} < \sqrt{75} < \underline{\hspace{1.5cm}}$$

$\sqrt{75}$ is between $\underline{\hspace{1cm}}$ and $\underline{\hspace{1cm}}$,

but closer to $\underline{\hspace{1cm}}$.

**7** $\sqrt{10}$

$$\underline{\hspace{1.5cm}} < \sqrt{10} < \underline{\hspace{1.5cm}}$$

$\sqrt{10}$ is between $\underline{\hspace{1cm}}$ and $\underline{\hspace{1cm}}$,

but closer to $\underline{\hspace{1cm}}$.

**8** $\sqrt{50}$ is between $\underline{\hspace{1cm}}$ and $\underline{\hspace{1cm}}$,

but closer to $\underline{\hspace{1cm}}$.

**9** $\sqrt{65}$ is between $\underline{\hspace{1cm}}$ and $\underline{\hspace{1cm}}$,

but closer to $\underline{\hspace{1cm}}$.

**10** $\sqrt{27}$ is between $\underline{\hspace{1cm}}$ and $\underline{\hspace{1cm}}$,

but closer to $\underline{\hspace{1cm}}$.

**11** $\sqrt{119}$ is between $\underline{\hspace{1cm}}$ and $\underline{\hspace{1cm}}$,

but closer to $\underline{\hspace{1cm}}$.

## Step by Step Problem-Solving Practice

**Solve.**

12 **SCHOOL** Lydia's math teacher made everything in her class about numbers. One day she returned quizzes with scores written as irrational numbers. Her teacher told the class that quiz scores would be recorded as the integer closest to the estimated value of the radical. Lydia's quiz had $\sqrt{128}$ written on it. What was the recorded score of Lydia's quiz?

Complete the inequality.

_____ $< \sqrt{128} <$ _____

Find the square root of each perfect square.

_____ $=$ _____          _____ $=$ _____

Because 128 is closer to _____ than _____,

_____ is closer to _____.

Lydia's quiz score was recorded as _____.

Check off each step.

_____ **Understand: I underlined key words.**

_____ **Plan: To solve the problem, I will** _____.

_____ **Solve: The answer is** _____.

_____ **Check: I checked my answer by** _____.

## ▶ Skills, Concepts, and Problem Solving

**Circle the word that classifies each number.**

13  $\dfrac{2}{3}$          rational          irrational

14  0.5          rational          irrational

15  $\sqrt{60}$          rational          irrational

16  $\sqrt{225}$          rational          irrational

GO ON

**Write each number below in the appropriate column.**

17  $\sqrt{6}, \dfrac{6}{6}, \sqrt{87}, \sqrt{64}, 0.5214, \sqrt{105}, 1\dfrac{1}{4}, \sqrt{1}, 0.\overline{16}, 18.5, \pi, 2.04\overline{5}$

| Rational Numbers | Irrational Numbers |
|---|---|
|  |  |
|  |  |

**Estimate the value of each number.**

18  $\sqrt{20}$ is between ___ and ___, closer to ___

19  $\sqrt{5}$ is between ___ and ___, closer to ___

20  $\sqrt{35}$ is between ___ and ___, closer to ___

21  $\sqrt{130}$ is between ___ and ___, closer to ___

**Solve.**

22  **BASEBALL**   Derek had 4 hits out of 15 times at bat. Write a fraction showing the number of hits over the number of times at bat. Is this a rational or irrational number?

_____

23  **AREA**   Lazaro knows the length of a side of a square is the square root of the area of the square. What is the length of one of the sides of the square shown?

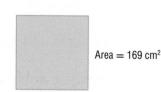

Area = 169 cm²

_____

**Vocabulary Check   Write the vocabulary word that completes each sentence.**

24  A(n) _____ can be written as a ratio of integers, where the denominator cannot equal 0.

25  Numbers that cannot be written as a ratio of two integers are called

_____ numbers.

26  **Reflect**   Explain the difference between a terminating decimal and a repeating decimal. Give an example of each type of decimal.

_____

_____

_____

# Fractions and Decimals

## KEY Concept

Fractions and decimals are different ways to show the same value.

$$\frac{2}{5} \quad = \quad 0.4$$

fraction → decimal

Divide the numerator by the denominator.

$$\frac{2}{5} = 2 \div 5 \rightarrow 5\overline{)2.0}$$
$$\phantom{5)}0.4$$
$$\phantom{5)}\underline{2.0}$$
$$\phantom{5)2.}0$$

decimal → fraction

Write the decimal as a fraction with a denominator that is a multiple of 10. Simplify the fraction.

$$0.4 = \frac{4}{10} = \frac{4 \div 2}{10 \div 2} = \frac{2}{5}$$

VOCABULARY

**decimal**
a number that can represent whole numbers and fractions; a decimal point separates the whole number from the fraction

**denominator**
the number below the bar in a fraction that tells how many equal parts in the whole or the set

**fraction**
a number that represents part of a whole or part of a set

**numerator**
the number above the bar in a fraction that tells how many equal parts are being used

If you use a calculator, you need to know the division problem so that you can enter the digits in the calculator correctly.

## Example 1

Write $\frac{1}{8}$ as a decimal.

1. Write the fraction as a division problem.

   $1 \div 8$

2. Use your calculator.

   [ 1 ] [ ÷ ] [ 8 ] [ = ]  0.125

## YOUR TURN!

Write $\frac{13}{25}$ as a decimal.

1. Write the fraction as a division problem.

   _____

2. Use your calculator.

   ▢ ▢ ▢ ▢ ▢ ▢  _____

GO ON

## Example 2

Write $\frac{3}{8}$ as a decimal.

1. Write the division problem.

   $8\overline{)3}$

2. Divide.

   $$\begin{array}{r} 0.375 \\ 8\overline{)3.000} \\ -2.4\phantom{00} \\ \hline 60\phantom{0} \\ -56\phantom{0} \\ \hline 40 \\ -40 \\ \hline 0 \end{array}$$

## YOUR TURN!

Write $\frac{7}{28}$ as a decimal.

1. Write the division problem.

   _____

2. Divide.

   $28\overline{)7.00}$

   _____

   _____

## Example 3

Write 0.036 as a fraction in simplest form.

1. Count the number of decimal places. What multiple of 10 is the denominator?
   1000

2. What is the numerator?

   36

3. Write the fraction.
   $\dfrac{36}{1000}$

4. Simplify. Divide the numerator and denominator by the Greatest Common Factor (GCF).

   $$\frac{36 \div 4}{1000 \div 4} = \frac{9}{250}$$

## YOUR TURN!

Write 0.85 as a fraction in simplest form.

1. Count the number of decimal places. What multiple of 10 is the denominator?

   Multiples of 10 are 10, 100, 1000, …

   _____

2. What is the numerator?

   _____

3. Write the fraction.

   _____

4. Simplify. Divide the numerator and denominator by the Greatest Common Factor (GCF).

   $$\frac{\Box \div \Box}{\Box \div \Box} = \frac{\Box}{\Box}$$

**Guided Practice**

**Write each fraction as a decimal.**

**1** $\dfrac{33}{50}$

$50\overline{)33.00}$

**2** $\dfrac{2}{5}$

$5\overline{)2.0}$

**3** $\dfrac{20}{32}$

$32\overline{)20.000}$

**4** $\dfrac{21}{25}$

$25\overline{)21.00}$

*Step* by *Step* **Practice**

**5** Write 0.72 as a fraction in simplest form.

**Step 1** Count the number of decimal places. What multiple of 10 is the denominator?

_____

**Step 2** What is the numerator?

_____

**Step 3** Write the fraction.

_____

**Step 4** Simplify. Divide the numerator and denominator by the GCF.

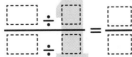

GO ON

**Write each decimal as a fraction in simplest form. Divide by the GCF.**

**6**   0.02

$$\frac{\boxed{\phantom{0}} \div \boxed{\phantom{0}}}{\boxed{\phantom{0}} \div \boxed{\phantom{0}}} = \frac{\boxed{\phantom{0}}}{\boxed{\phantom{0}}}$$

**7**   0.375

$$\frac{\boxed{\phantom{0}} \div \boxed{\phantom{0}}}{\boxed{\phantom{0}} \div \boxed{\phantom{0}}} = \frac{\boxed{\phantom{0}}}{\boxed{\phantom{0}}}$$

**8**   0.16

$$\frac{\boxed{\phantom{0}} \div \boxed{\phantom{0}}}{\boxed{\phantom{0}} \div \boxed{\phantom{0}}} = \frac{\boxed{\phantom{0}}}{\boxed{\phantom{0}}}$$

**9**   0.005

$$\frac{\boxed{\phantom{0}} \div \boxed{\phantom{0}}}{\boxed{\phantom{0}} \div \boxed{\phantom{0}}} = \frac{\boxed{\phantom{0}}}{\boxed{\phantom{0}}}$$

## Step by Step Problem-Solving Practice

**Solve.**

**10**   **BASKETBALL**   Luke made 9 out of 20 free throws during basketball practice. Write the fraction and decimal that shows the number of free throws Luke made.

$$20\overline{)9.00}$$

_____

Check off each step.

_____ **Understand: I underlined key words.**

_____ **Plan: To solve the problem, I will** _____.

_____ **Solve: The answer is** _____.

_____ **Check: I checked my answer by** _____.

# Skills, Concepts, and Problem Solving

**Write each fraction as a decimal or each decimal as a fraction in simplest form.**

**11** $\dfrac{5}{8}$ _____

**12** $\dfrac{13}{20}$ _____

**13** 0.48 _____

**14** 0.6 _____

**15** 0.05 _____

**16** 0.86 _____

**17** $0.\overline{3}$ _____

**18** $\dfrac{1}{9}$ _____

**19** 0.59 _____

**Solve. Write your answers in simplest form.**

**20** **TESTING** Mr. Arias has 64 students in his math classes. On the last test, 16 students earned a perfect score. Write the fraction and decimal that shows the number of students who earned perfect scores.

_____

**21** **SURVEY** The school newspaper took a survey regarding the kind of school pizza the students liked the most. The results are given in the table. Write the fraction and decimal showing the number of students who like mushroom pizza the best.

_____

| Type of Pizza | Number of Students |
|---|---|
| Cheese | 44 |
| Pepperoni | 58 |
| Mushroom | 18 |

**Vocabulary Check** **Write the vocabulary word that completes each sentence.**

**22** The number above the bar in a fraction is the _____.

**23** A(n) _____ can represent a whole number or a fraction.

**24** **Reflect** When performing a long division problem, describe how you can tell that your answer is going to be a repeating decimal.

_____

_____

_____

# Progress Check 1 (Lessons 2-1 and 2-2)

**Circle the word that classifies each number.**

1  $\frac{1}{8}$        rational    irrational

2  $0.\overline{3}$        rational    irrational

3  $\sqrt{81}$        rational    irrational

4  $\sqrt{3}$        rational    irrational

**Estimate the value of each number.**

5  $\sqrt{26}$ is between _____ and _____, but closer to _____.

6  $\sqrt{47}$ is between _____ and _____, but closer to _____.

7  $\sqrt{114}$ is between _____ and _____, but closer to _____.

8  $\sqrt{205}$ is between _____ and _____, but closer to _____.

**Write each fraction as a decimal or each decimal as a fraction in simplest form.**

9  $\frac{5}{4}$ _____

10  $\frac{11}{20}$ _____

11  0.33 _____

12  0.8 _____

13  0.011 _____

14  0.12 _____

15  0.5 _____

16  $\frac{4}{9}$ _____

17  1.5 _____

**Solve.**

18  **GAMES**  Seni has won the last 12 out of 15 card games he has played on his computer. Write the fraction in simplest form. Then write the decimal that shows his record of winning.

_____

19  **COOKING**  Tyna makes a spicy chili. For each 3 cups of chili, she uses 2 habanera peppers. Write a fraction showing the ratio of peppers to cups of chili. Is this a rational or irrational number?

_____

# Decimals and Percents

## KEY Concept

A **percent** is a ratio that compares a number to 100. Percent means "per hundred."

Decimals and percents are different ways to show the same value.

$$\frac{67}{100}$$

67 out of 100

0.67

67 hundredths

decimal → percent    Multiply the decimal by 100.
                     Write a percent symbol.

$$0.67 \times 100 = 67\%$$

percent → decimal    Remove the percent symbol
                     Divide the decimal by 100.

$$67\% \div 100 = 0.67$$

$$\frac{67}{100} = 0.67 = 67\%$$

### VOCABULARY

**decimal**
  a number that can represent whole numbers and fractions; a decimal point separates the whole number from the fraction

**percent**
  a ratio that compares a number to 100

**ratio**
  a comparison of two numbers by division

To multiply by 100, move the decimal point right two places. To divide by 100, move the decimal point left two places.

## Example 1

**Write 0.24 as a percent.**

1. Multiply 0.24 by 100.

   $$0.24 \times 100 = 24$$

2. The decimal point moved two places to the right.

3. Write a percent symbol.

   24%

### YOUR TURN!

**Write 1.37 as a percent.**

1. _____ 1.37 by _____.

   _____ × _____ = _____

2. The decimal point moved _____ places

   to the _____.

3. Write a percent symbol.

   _____

GO ON

## Example 2

**Write 2% as a decimal.**

1. Remove the percent symbol.

2. Divide 2 by 100.

> Add a zero to move the decimal.

$02 \div 100 = 0.02$

3. The decimal point moved two places to the left.

### YOUR TURN!

**Write 15% as a decimal.**

1. Remove the percent symbol.

2. _____ 15 by _____.

_____ ÷ _____ = _____

3. The decimal point moved _____ places

to the _____.

## ▶ Guided Practice

**Write each decimal as a percent.**

**1** 0.38

_____ × 100 = _____

Write a percent symbol.  _____

**2** 0.7

_____ × _____ = _____

Write a percent symbol.  _____

**3** 0.33

_____ × 100 = _____

**4** 6

_____ × 100 = _____

**5** 2.9

_____ × _____ = _____

**6** 0.05

_____ × _____ = _____

**7** 0.81

_____ × _____ = _____

**8** 1.3

_____ × _____ = _____

## Step by Step Practice

**9** Write 1.5% as a decimal.

**Step 1** _____ 1.5 by _____.

_____ ÷ _____ = _____

**Step 2** The decimal point moved _____ places

to the _____.

**Write each percent as a decimal.**

**10** 44%

_____ ÷ 100 = _____

**11** 3%

_____ ÷ 100 = _____

**12** 66%

_____ ÷ 100 = _____

**13** 9%

_____ ÷ 100 = _____

**14** 12.5%

_____ ÷ _____ = _____

**15** 254%

_____ ÷ _____ = _____

**16** 326%

_____ ÷ _____ = _____

**17** 29.4%

_____ ÷ _____ = _____

## Step by Step Problem-Solving Practice

**Solve.**

**18** SCHOOL    Elise's math exam had 50 problems on it. She was able to do 36 of them in 1 hour. What percent of the math problems did she complete?

_____    Change the ratio to a decimal.

_____    Change the decimal to a percent.

Check off each step.

_____ Understand: I underlined key words.

_____ Plan: To solve the problem, I will _____.

_____ Solve: The answer is _____.

_____ Check: I checked my answer by _____

_____.

GO ON

# Skills, Concepts, and Problem Solving

**Write each decimal as a percent or each percent as a decimal.**

**19** 0.82 _____

**20** 0.23 _____

**21** 3.2 _____

**22** 90% _____

**23** 11% _____

**24** 145% _____

**25** 14.92 _____

**26** 52.7% _____

**27** 3.18% _____

**28** 500% _____

**29** 0.049 _____

**30** 0.98 _____

**Solve.**

**31** JUICE   A bottle of juice says it has 15% real fruit juice. Write the amount of real fruit juice as a decimal.

_____

**32** BASKETBALL   Sarita made 0.44 of the shots she attempted in a basketball game. What percent of shots did Sarita make?

_____

**33** GRADES   The circle graph below shows how Mr. Escalante's grades are distributed in his social studies classes. Write the percent of students earning A's and B's as a decimal.

_____

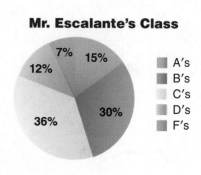

**Mr. Escalante's Class**

- A's
- B's
- C's
- D's
- F's

**Vocabulary Check**   **Write the vocabulary word that completes each sentence.**

**34** A(n) _____ is a comparison of two numbers by division.

**35** A(n) _____ is a ratio that compares a number to 100.

**36** **Reflect**   Write a percent and a decimal that show the same amount. Explain why 0.25 is not the same as 2.5%.

_____

_____

STOP

# Fractions and Percents

## KEY Concept

Fractions and percents are different ways to show the same value. There are different ways to convert between them.

percent → fraction    Remove the percent symbol. Write a fraction with a denominator of 100. Simplify.

$$25\% \rightarrow \frac{25}{100} = \frac{25 \div 25}{100 \div 25} = \frac{1}{4}$$

fraction → percent    Write and solve a proportion.

$$\frac{5}{8} = \frac{x}{100}$$

5 out of 8 is the same as $x$ out of 100.

$$500 = 8x$$    Cross multiply.

$$x = 62.5$$    Divide each side of the equation by 8.

$$\frac{5}{8} = 62.5\%$$

fraction → percent    Divide, and then multiply by 100.

$$\frac{5}{8} = 5 \div 8 \rightarrow 8 \overline{)5.000} \;\; ^{0.625}$$

$$0.625 \times 100 = 62.5\%$$

Copyright © Glencoe/McGraw-Hill, a division of The McGraw-Hill Companies, Inc.

### VOCABULARY

**denominator**
the number below the bar in a fraction that tells how many equal parts in the whole or the set

**fraction**
a number that represents part of a whole or set

**percent**
a ratio that compares a number to 100

Memorize commonly used fractions and percents, such as $\frac{1}{4} = 25\%$.

## Example 1

**Write 65% as a fraction.**

1. Write a fraction with a denominator of 100.

$$65\% = \frac{65}{100}$$

2. Simplify.

$$\frac{65 \div 5}{100 \div 5} = \frac{13}{20}$$

## YOUR TURN!

**Write 120% as a fraction.**

1. Write a fraction with a denominator of 100.

$$120\% = \frac{\boxed{\phantom{x}}}{100}$$

2. Simplify.

$$\frac{\boxed{\phantom{x}} \div \boxed{\phantom{x}}}{100 \div \boxed{\phantom{x}}} = \underline{\phantom{xx}} = \underline{\phantom{xx}}$$

**GO ON**

## Example 2

Write $\frac{12}{5}$ as a percent.

1. Write a proportion.

$$\frac{12}{5} = \frac{x}{100}$$

2. Cross multiply and solve for $x$.

$$12 \cdot 100 = 5 \cdot x$$

$$1{,}200 = 5x$$

$$240 = x$$

3. Write the percent.

$$\frac{12}{5} = 240\%$$

**YOUR TURN!**

Write $\frac{8}{20}$ as a percent.

1. Write a proportion.

$$\frac{\boxed{\phantom{x}}}{\boxed{\phantom{x}}} = \frac{x}{100}$$

2. Cross multiply and solve for $x$.

$$\underline{\hspace{3cm}} = \underline{\hspace{3cm}}$$

$$\underline{\hspace{2cm}} = \underline{\hspace{2cm}}$$

$$\underline{\hspace{2cm}} = \underline{\hspace{2cm}}$$

3. Write the percent.

$$\frac{8}{20} = \underline{\hspace{2cm}}$$

 **Guided Practice**

**Write each percent as a fraction.**

**1** 30%

$$30\% = \frac{\boxed{\phantom{x}}}{100}$$

$$\frac{\boxed{\phantom{x}} \div \boxed{\phantom{x}}}{100 \div \boxed{\phantom{x}}} = \underline{\hspace{2cm}}$$

**2** 65%

$$65\% = \frac{\boxed{\phantom{x}}}{100}$$

$$\frac{\boxed{\phantom{x}} \div \boxed{\phantom{x}}}{100 \div \boxed{\phantom{x}}} = \underline{\hspace{2cm}}$$

**Step** (by) **Step Practice**

**3** Write $\frac{9}{4}$ as a percent.

**Step 1** Write a proportion. The fraction for an unknown percent is $\frac{x}{100}$.

$$\frac{\boxed{\phantom{x}}}{\boxed{\phantom{x}}} = \frac{x}{100}$$

**Step 2** Cross multiply and solve for $x$.

$$\underline{\hspace{1.5cm}} \cdot \underline{\hspace{1.5cm}} = \underline{\hspace{1.5cm}} \cdot \underline{\hspace{1.5cm}}$$

$$\underline{\hspace{1.5cm}} = \underline{\hspace{1.5cm}}$$

$$x = \underline{\hspace{1.5cm}}$$

**Step 3** Write the percent.

$$\frac{9}{4} = \underline{\hspace{1.5cm}}$$

**Write each fraction as a percent.**

**4** $\dfrac{3}{15}$

_____ = _____

_____ • _____ = _____ • _____

_____ = _____

$x =$ _____

$\dfrac{3}{15} =$ _____%

**5** $\dfrac{7}{20}$

_____ = _____

_____ • _____ = _____ • _____

_____ = _____

$x =$ _____

$\dfrac{7}{20} =$ _____%

**Write each fraction as a percent or each percent as a fraction in simplest form.**

**6** 66%

$66\% = \dfrac{\boxed{\phantom{00}}}{100}$

$\dfrac{\boxed{\phantom{0}} \div \boxed{\phantom{0}}}{\boxed{\phantom{0}} \div \boxed{\phantom{0}}} =$ _____

**7** $\dfrac{7}{8}$

_____ = _____

_____ = _____

$x =$ _____

_____%

**8** 17%

$\dfrac{\boxed{\phantom{00}}}{\boxed{\phantom{00}}}$

**9** 95%

$\dfrac{\boxed{\phantom{0}}}{\boxed{\phantom{0}}}$

$\dfrac{\boxed{\phantom{0}} \div \boxed{\phantom{0}}}{\boxed{\phantom{0}} \div \boxed{\phantom{0}}} =$ _____

**10** $\dfrac{11}{4}$

_____ = _____

_____ = _____

$x =$ _____

_____%

**11** $\dfrac{3}{16}$

_____ = _____

_____ = _____

$x =$ _____

_____%

GO ON

**Solve.**

12  **PIZZA**   Dario and Tyrone ordered a large pizza. Together they ate 9 out of the 12 slices. What percent of the pizza did Dario and Tyrone eat together?

Write a proportion using the ratio.   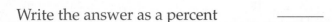 \_\_\_\_ = \_\_\_\_

Solve for *x*.   \_\_\_\_ = \_\_\_\_

\_\_\_\_ = \_\_\_\_

Write the answer as a percent   \_\_\_\_

Check off each step.

_____ **Understand: I underlined key words.**

_____ **Plan: To solve the problem, I will** _____.

_____ **Solve: The answer is** _____.

_____ **Check: I checked my answer by** _____.

 ## Skills, Concepts, and Problem Solving

**Write each fraction as a percent or each percent as a fraction in simplest form.**

13  12%

_____

14  85%

_____

15  170%

_____

16  17%

_____

17  $\dfrac{5}{8}$

_____

18  $\dfrac{5}{4}$

_____

19  $\dfrac{17}{25}$

_____

20  $\dfrac{18}{24}$

_____

**Solve. Write the answer in simplest form.**

**21** **EXERCISE** Ivana runs 1 mile each morning. So far this morning, she has run $\frac{33}{75}$ mile. What percent of 1 mile has she run?

_____

**22** **WORK** Ava has completed $\frac{3}{8}$ of a new project she has at work. What percent of her new project has she completed?

_____

**23** **TRAVEL** Deon is packing for a trip. He has completed 80% of his packing. Write a fraction to show how much Deon has packed for his trip.

_____

**24** **RESTAURANT** Namid went to a restaurant with some friends. She looks at the menu below. What percent of the items on the menu are chicken?

_____

*Menu*

Hamburger. . . . . . . . . . . . . $ 5.50
Chicken Patty . . . . . . . . . . . $ 5.25
Steak Sandwich . . . . . . . . . $ 7.50
Pork BBQ . . . . . . . . . . . . . . $ 6.00
Chicken Nachos . . . . . . . . . $ 5.50

**Vocabulary Check** **Write the vocabulary word that completes each sentence.**

**25** A(n) _____ is a ratio that compares a number to 100.

**26** A number that represents part of a whole is a(n) _____.

**27** **Reflect** What type of fraction represents percents greater than 100%? What does it mean to have a percent greater than 100%?

_____

_____

**Write each decimal as a percent and each percent as a decimal.**

**1** 0.22 _____

**2** 0.18 _____

**3** 0.76 _____

**4** 125% _____

**5** 10% _____

**6** 8% _____

**7** 3.5 _____

**8** 14.2% _____

**9** 6.55% _____

**10** 700% _____

**11** 0.088 _____

**12** 0.9 _____

**Write each fraction as a decimal or each decimal as a fraction in simplest form.**

**13** 64%

_____

**14** 5%

_____

**15** 220%

_____

**16** 97%

_____

**17** $\frac{5}{16}$

_____

**18** $\frac{50}{10}$

_____

**Solve. Write the answer in simplest form.**

**19** **EXERCISE**   David has completed $\frac{5}{8}$ of the exercises his trainer asked him to do. What percent of his exercises has he completed?

_____

**20** **HYGIENE**   A new type of body lotion claims you will see a 33% improvement in your skin's texture in 2 weeks. Write the amount of improvement you will see as a decimal.

_____

# Simplify Square Roots

## KEY Concept

The number 9 is a **perfect square** because 9 is the result of multiplying the factor 3 two times.

$$3^2 = 3 \cdot 3 = 9$$

The inverse of squaring a number is finding the **square root** of a number. The symbol $\sqrt{\phantom{x}}$ denotes taking the square root.

$$\pm\sqrt{9} = \pm 3, \text{ because } 3^2 = 9 \text{ and } (-3)^2 = 9.$$

A radical is not simplified if there are any perfect square factors of the number under the radical sign. Consider $\sqrt{18}$. The number under the radical sign, 18, has factors of 2 and 9. Since 9 is a perfect square, $\sqrt{18}$, is not simplified.

$$\sqrt{18} = \sqrt{9 \cdot 2} = \sqrt{9} \cdot \sqrt{2} = 3\sqrt{2}$$

| **Not Simplified** | | **Simplified** | |
|:---:|:---:|:---:|:---:|
| $\sqrt{45}$ | $\sqrt{24}$ | $\sqrt{3}$ | $\sqrt{30}$ |

$\sqrt{45}$ is not simplified because 9 is a factor of 45.

$$\sqrt{45} = \sqrt{9.5} = \sqrt{9} \cdot \sqrt{5} = 3\sqrt{5}$$

$\sqrt{24}$ is not simplified because 4 is a factor of 24.

$$\sqrt{24} = \sqrt{4.6} = \sqrt{4} \cdot \sqrt{6} = 2\sqrt{6}$$

## VOCABULARY

**perfect square**
a number with a root that is rational

**radical sign**
the symbol $\sqrt{\phantom{x}}$ used to indicate a nonnegative square root

**square number**
the product of a number multiplied by itself

**square root**
one of the two equal factors of a number

$\sqrt{16}$ represents the positive square root of 16, or 4. $-\sqrt{16}$ represents the negative square root of 16, or $-4$. $\pm\sqrt{16}$ represents both the positive and negative square root of 16.

## Example 1

**Simplify $\pm\sqrt{36}$.**

1. What number(s) times itself equals 36?

$$6^2 = 36 \qquad (-6)^2 = 36$$

2. The sign before the radical means both the positive and negative square root.

$$\pm\sqrt{36} = \pm 6$$

## YOUR TURN!

**Simplify $-\sqrt{49}$.**

1. What number(s) times itself equals _____?

$$\boxed{\phantom{x}}^2 = 49 \qquad (\boxed{\phantom{x}})^2 = 49$$

2. The sign before the radical means the

_____ square root.

$$-\sqrt{49} = \underline{\phantom{xx}}$$

GO ON

## Example 2

**Simplify $\sqrt{40}$.**

1. 40 is divisible by the perfect square 4.

$$40 = 4 \cdot 10$$

2. Write $\sqrt{40}$ as a product.

$$\sqrt{40} = \sqrt{4} \cdot \sqrt{10}$$
$$= 2\sqrt{10}$$

**YOUR TURN!**

List the factors of 75. Are any of the factors perfect squares?

**Simplify $\sqrt{75}$.**

1. 75 is divisible by the perfect square _____.

$$75 = \underline{\phantom{xx}} \cdot \underline{\phantom{xx}}$$

2. Write $\sqrt{75}$ as a product.

$$\sqrt{75} = \sqrt{\boxed{\phantom{x}}} \cdot \sqrt{\boxed{\phantom{x}}}$$
$$= \boxed{\phantom{x}}\sqrt{\boxed{\phantom{x}}}$$

## ▶ Guided Practice

**Simplify each square root.**

**1** $\sqrt{225}$

$$\boxed{\phantom{x}}^2 = 225$$

$$\sqrt{225} = \underline{\phantom{xxx}}$$

**2** $\pm\sqrt{121}$

$$\boxed{\phantom{x}}^2 = 121$$

$$\pm\sqrt{121} = \underline{\phantom{xxx}}$$

**3** $-\sqrt{16}$

$$\boxed{\phantom{x}}^2 = 16$$

$$-\sqrt{16} = \underline{\phantom{xxx}}$$

**4** $\sqrt{49} = \underline{\phantom{xxx}}$

**5** $\sqrt{81} = \underline{\phantom{xxx}}$

**6** $-\sqrt{100} = \underline{\phantom{xxx}}$

**7** $-\sqrt{25} = \underline{\phantom{xxx}}$

**8** $\pm\sqrt{36} = \underline{\phantom{xxx}}$

**9** $\sqrt{196} = \underline{\phantom{xxx}}$

## Step by Step Practice

**10** Simplify $\sqrt{153}$.

**Step 1** 153 is divisible by the perfect square _____.

$$153 = \underline{\phantom{xx}} \cdot \underline{\phantom{xx}}$$

**Step 2** Write $\sqrt{153}$ as a product.

$$\sqrt{153} = \sqrt{\boxed{\phantom{x}}} \cdot \sqrt{\boxed{\phantom{x}}} = \boxed{\phantom{x}}\sqrt{\boxed{\phantom{x}}}$$

**Simplify each square root.**

**11** $\sqrt{50}$

$50 = $ _____ • _____

$\sqrt{50} = $ _____

**12** $\pm\sqrt{96}$

$96 = $ _____ • _____

$\pm\sqrt{96} = $ _____

**13** $-\sqrt{20}$

$20 = $ _____ • _____

$-\sqrt{20} = $ _____

**14** $-\sqrt{45} = $ _____

**15** $\pm\sqrt{128} = $ _____

**16** $-\sqrt{400} = $ _____

**17** $\sqrt{294} = $ _____

**18** $-\sqrt{152} = $ _____

**19** $\sqrt{72} = $ _____

**20** $\sqrt{121} = $ _____

**21** $-\sqrt{243} = $ _____

**22** $\pm\sqrt{108} = $ _____

## Step by Step *Problem-Solving Practice*

**Solve.**

**23** **AREA**   Kurt is fencing in his yard. The shape of his yard is a square. Kurt knows that the area of his yard is 196 square yards. What is the length of each side of his yard?

Kurt's yard is in the shape of a _____.

The length of each side is equal to the _____ of 196.

_____ = _____

_____ yards

Check off each step.

_____ Understand: I underlined key words.

_____ Plan: To solve the problem, I will _____.

_____ Solve: The answer is _____.

_____ Check: I checked my answer by _____.

**GO ON**

## ▶ Skills, Concepts, and Problem Solving

**Simplify each square root.**

**24** $\sqrt{4} =$ _____

**25** $\pm\sqrt{12} =$ _____

**26** $-\sqrt{25} =$ _____

**27** $\sqrt{80} =$ _____

**28** $\sqrt{24} =$ _____

**29** $\sqrt{49} =$ _____

**30** $-\sqrt{400} =$ _____

**31** $\pm\sqrt{90} =$ _____

**32** $-\sqrt{162} =$ _____

**Solve.**

**33** **COMPUTERS** Marlee is looking to buy a new computer screen. The computer screen is a square. The area of the screen she likes is 169 square inches. What is the length of each side of the screen?

_____

**34** **MEDIA** The width of a DVD case is the same as the width of a DVD. If the DVD case is a square with an area of 144 square centimeters, what is the width of the DVD?

_____

**35** **AREA** What is the length of each side in the square to the right?

_____

Area = 98 cm²

**Vocabulary Check** **Write the vocabulary word that completes each sentence.**

**36** A(n) _____ is used to indicate a square root.

**37** A(n) _____ is one of the two equal factors of a number.

**38** The product of a number multiplied by itself is a(n) _____.

**39** **Reflect** Can you simplify the square root of a prime number? Explain your answer and give an example.

_____

_____

_____

**STOP**

# Compare and Order Real Numbers

## KEY Concept

The set of **real numbers** is all the rational and the irrational numbers.

**Real Numbers**

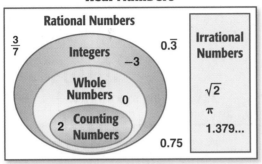

To compare real numbers, write rational numbers as decimals and approximate irrational numbers as decimals. Use the symbols <, >, or =.

$$-\sqrt{4} < 0.75 < \sqrt{16}$$

Copyright © Glencoe/McGraw-Hill, a division of The McGraw-Hill Companies, Inc.

### VOCABULARY

**irrational number**
  any number that cannot be written as a fraction $\frac{a}{b}$, where $a$ and $b$ are integers and $b \neq 0$

**rational number**
  any number that can be written as a fraction $\frac{a}{b}$, where $a$ and $b$ are integers and $b \neq 0$

**real numbers**
  the set of rational and irrational numbers

To compare real numbers, compare place values or think of each number's placement on a number line.

## Example 1

**Identify all sets to which 2.35 belongs.**

1. 2.35 is a real number.

2. Because the number 2.35 terminates or repeats. It is a rational number.

3. The number 2.35 is a decimal. It is not an integer, whole number, or counting number.

4. The number 2.35 belongs to the sets of real and rational numbers.

## YOUR TURN!

**Identify all sets to which 2.7183… belongs.**

1. 2.7183… is a _____ number.

2. Because the number 2.7183… _____ terminate or repeat it is a(n) _____ number.

3. The number 2.7183… is a(n) _____ number. It _____ an integer, whole number, or counting number.

4. The number 2.7183… belongs to the sets of _____ numbers.

GO ON

## Example 2

Use >, <, or = to compare $-9$ and $-\dfrac{19}{6}$.

1. Write each number in decimal form.

$$-9 = -9.0 \qquad -\dfrac{19}{6} = -3.1\overline{6}$$

2. On a number line, $-\dfrac{19}{6}$ is farther right.

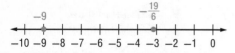

3. Write an inequality. $-\dfrac{19}{6} > -9$

---

**YOUR TURN!**

Use >, <, or = to compare $\dfrac{1}{4}$ and $\dfrac{1}{3}$.

1. Write each number in decimal form.

$$\dfrac{1}{4} = \underline{\hspace{2cm}} \qquad \dfrac{1}{3} = \underline{\hspace{2cm}}$$

2. On a number line, _____ is farther right.

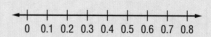

3. Write an inequality. $\dfrac{1}{4} \bigcirc \dfrac{1}{3}$

---

## Example 3

**Order the numbers from least to greatest.**

$$3\dfrac{3}{4}, \ \sqrt{81}, \ -8.\overline{6}, \ -\sqrt{49}$$

1. Write each number as a decimal.

$$3\dfrac{3}{4} = 3.75 \qquad -8.\overline{6} = -8.666\ldots$$
$$\sqrt{81} = 9.0 \qquad -\sqrt{49} = -7.0$$

2. Compare the decimals.

3. Write in order from least to greatest.

$$-8.\overline{6}, \ -\sqrt{49}, \ 3\dfrac{3}{4}, \ \sqrt{81}$$

---

**YOUR TURN!**

**Order the numbers from least to greatest.**

$$-\sqrt{64}, \ 8.\overline{43}, \ -6, \ \sqrt{4}$$

1. Write each number as a decimal.

$$-\sqrt{64} = \underline{\hspace{1.5cm}} \qquad -6 = \underline{\hspace{1.5cm}}$$
$$8.\overline{43} = \underline{\hspace{1.5cm}} \qquad \sqrt{4} = \underline{\hspace{1.5cm}}$$

2. Compare the decimals.

3. Write in order from least to greatest.

$$\underline{\hspace{5cm}}$$

---

 **Guided Practice**

**Identify all sets to which each number belongs.**

**1** $-5.34$

_____

_____

**2** $\sqrt{169}$

_____

_____

**3** $\dfrac{2}{3}$

_____

**4** $\sqrt{2}$

_____

**5** Order the numbers from least to greatest. $\quad \frac{1}{8}, -2.45, -\sqrt{16}, \frac{9}{4}$

**Step 1** Write each number as a decimal.

$$3\frac{1}{8} = \underline{\hspace{2cm}} \qquad -2.45 \qquad -\sqrt{16} = \underline{\hspace{2cm}} \qquad \frac{9}{4} = \underline{\hspace{2cm}}$$

**Step 2** Compare the decimals. $\quad\underline{\hspace{1.5cm}} < \underline{\hspace{1.5cm}} \qquad \underline{\hspace{1.5cm}} < \underline{\hspace{1.5cm}}$

**Step 3** Write in order from least to greatest. $\quad\underline{\hspace{1.2cm}}, \underline{\hspace{1.2cm}}, \underline{\hspace{1.2cm}}, \underline{\hspace{1.2cm}}$

**Order the numbers from least to greatest.**

**6** $2\frac{1}{3}, -0.5, \sqrt{9}, 0.\overline{3}$

$2\frac{1}{3} = \underline{\hspace{2cm}} \qquad\qquad -0.5 = \underline{\hspace{2cm}}$

$\sqrt{9} = \underline{\hspace{2cm}} \qquad\qquad 0.\overline{3} = \underline{\hspace{2cm}}$

$\underline{\hspace{5cm}}$

**7** $\sqrt{36}, 5.\overline{6}, -1.25, -\sqrt{1}$

$\sqrt{36} = \underline{\hspace{2cm}} \qquad\qquad 5.\overline{6} = \underline{\hspace{2cm}}$

$-1.25 = \underline{\hspace{2cm}} \qquad\qquad -\sqrt{1} = \underline{\hspace{2cm}}$

$\underline{\hspace{5cm}}$

**Solve.**

**8** QUALITY CONTROL Melanie is a quality control manager. The following numbers are error readings for a machine. Order the error readings from least to greatest.

$$\sqrt{25}, 1\frac{1}{3}, \frac{27}{5}, 3.\overline{18}$$

$\sqrt{25} = \underline{\hspace{1.5cm}} \qquad 1\frac{1}{3} = \underline{\hspace{1.5cm}} \qquad \frac{27}{5} = \underline{\hspace{1.5cm}} \qquad 3.\overline{18} = \underline{\hspace{1.5cm}}$

$\underline{\hspace{5cm}}$

Check off each step.

$\underline{\hspace{1.5cm}}$ Understand: I underlined key words.

$\underline{\hspace{1.5cm}}$ Plan: To solve the problem, I will $\underline{\hspace{6cm}}$.

$\underline{\hspace{1.5cm}}$ Solve: The answer is $\underline{\hspace{6cm}}$.

$\underline{\hspace{1.5cm}}$ Check: I checked my answer by $\underline{\hspace{6cm}}$.

GO ON

# Skills, Concepts, and Problem Solving

**Identify all sets to which the number belongs.**

**9** $\sqrt{10}$

_____

**10** $-9$

_____

**Order the numbers from least to greatest.**

**11** $-3.12, -\sqrt{9}, -\frac{7}{2}, -5$

_____

**12** $\sqrt{75}, 7.\overline{38}, \sqrt{100}, 12.12$

_____

**13** $-6\frac{1}{3}, -\sqrt{36}, -\frac{8}{3}, \sqrt{16}$

_____

**14** $-\sqrt{121}, -10.\overline{66}, -9.85, -10\frac{1}{4}$

_____

**Solve.**

**15** **MEASUREMENT** Avery has measured the width of four items and recorded her measurements in the table. She puts them in order by width from least to greatest. What order are the items?

_____

| ITEM | WIDTH |
|--------|--------|
| stamp | 2 mm |
| marble | $1\frac{1}{2}$ mm |
| eraser | $2\frac{1}{4}$ mm |
| penny | 1.67 mm |

**16** **WOODWORKING** Landon measured 4 pieces of wood for a project he was building. The lengths of the pieces of wood were 2.25 in., $2\frac{2}{3}$ in., 4 in., and $2\frac{3}{8}$ in. If Landon wants to put the pieces in order from shortest to longest, what would be the order?

_____

**Vocabulary Check** **Write the vocabulary word that completes each sentence.**

**17** The set of all rational and irrational numbers is the _____.

**18** A(n) _____ can be written as a fraction of two integers.

**19** **Reflect** Is it always necessary to simplify a non-perfect square root when you are putting numbers in order? Explain your answer.

_____

_____

**STOP**

**Simplify each square root.**

**1** $\sqrt{4}$ _____

**2** $\pm\sqrt{18}$ _____

**3** $-\sqrt{900}$ _____

**4** $\pm\sqrt{32}$ _____

**5** $\sqrt{125}$ _____

**6** $\sqrt{63}$ _____

**7** $-\sqrt{144}$ _____

**8** $\sqrt{98}$ _____

**9** $-\sqrt{99}$ _____

**Order each set of numbers from least to greatest.**

**10** $5\frac{2}{3}, \sqrt{25}, 6.1, 5\frac{3}{4}$

$5\frac{2}{3} =$ _____   $\sqrt{25} =$ _____

$6.1 =$ _____   $5\frac{3}{4} =$ _____

_____

**11** $-9.1, -2\frac{1}{4}, 2.\overline{1}, -\sqrt{81}$

_____

**Identify all sets to which each number belongs.**

**12** $-2.\overline{45}$

_____

_____

**13** $\sqrt{64}$

_____

_____

**Solve.**

**14** **MEASUREMENT**   Alice measured the width of the three windows in her living room and recorded her measurements in the table. She wants to put them in order by width from greatest to least. What order does she write the width of the windows?

_____

| Window | Width |
|--------|-------|
| North 1 | 60 inches |
| North 2 | 24.8 inches |
| West | 30.5 inches |

**15** **DECORATING**   Ralph needs a square frame for a picture that has an area of 144 square inches. What is the length of each side of the frame Ralph needs?

_____

**Circle the word that classifies each number.**

**1** $\pi$        rational    irrational      **2** $1.\overline{6}$      rational    irrational

**Estimate the value of each number.**

**3** $\sqrt{12}$ is between _____ and _____, but closer to _____.

**4** $\sqrt{72}$ is between _____ and _____, but closer to _____.

**5** $-\sqrt{30}$ is between _____ and _____, but closer to _____.

**Write each fraction as a decimal or each decimal as a fraction in simplest form.**

**6** $\dfrac{1}{16}$ _____       **7** $\dfrac{15}{40}$ _____       **8** 0.32 _____

**9** 0.4 _____       **10** 0.003 _____       **11** 0.89 _____

**Write each decimal as a percent or each percent as a decimal.**

**12** 0.41 _____       **13** 0.75 _____       **14** 4.5 _____

**15** 1% _____       **16** 13% _____       **17** 130% _____

**Identify all sets to which each number belongs.**

**18** −5                           **19** $\sqrt{10}$

_____      _____

_____      _____

**Write each percent as a fraction in simplest form.**

**20** 16%                          **21** 17%

_____                              _____

**22** 7%                            **23** 65%

_____                              _____

**Simplify each square root.**

**24** $\sqrt{25}$ _____

**25** $\sqrt{44}$ _____

**26** $-\sqrt{81}$ _____

**27** $\pm\sqrt{120}$ _____

**28** $\sqrt{54}$ _____

**29** $-\sqrt{121}$ _____

**Order each set of numbers from least to greatest.**

**30** $\frac{2}{3}, \sqrt{60}, 72.1, 5\frac{3}{4}, -\sqrt{36}$ _____

**31** $-1.25, -2\frac{1}{4}, 1.8, \sqrt{3}$ _____

**Solve.**

**32** **CLOTHES** Rochelle has four skirts she likes to wear. Her mother told her to give the second shortest skirt to charity. Rochelle's skirts are the following lengths: black, 24.5 in.; blue, 19.3 in.; tan, 18.$\overline{6}$ in.; and red 22.125 in. Which skirt will be given away?

_____

**33** **GARDENING** Mr. Green planted 3 dozen tulip bulbs in the fall. In the spring, 22 bloomed. Write the fraction and decimal that shows the number of tulips that bloomed.

_____

**Correct the Mistake**

**34** Jennifer classifies the number $-\sqrt{289}$ as a real number and an irrational number. Her teacher said she has one correct classification and one incorrect classification. Explain Jennifer's answer and correct her classifications.

_____

_____

_____

_____

# Chapter
## 3

# Equations and Inequalities

### How many yards were gained?

A football team gained 10 yards after running two plays. The first play resulted in a loss of 6 yards. You can write and solve an equation to find the total number of yards gained in the second play.

$$-6 + x = 10$$

**STEP 2 Preview**    Get ready for Chapter 3. Review these skills and compare them with what you will learn in this chapter.

| What You Know | What You Will Learn |
|---|---|

**What You Know**

You know how to evaluate expressions.

**Example:**

Evaluate the expression $x + 3$ if $x = 2$.

$x + 3 = 2 + 3$    Replace $x$ with 2.
$\phantom{x + 3} = 5$      Add.

**TRY IT!**

Evaluate each expression if $x = 4$ and $y = 5$.

**1**   $x + 6 =$ _____

**2**   $3 + y =$ _____

**3**   $y + 16 =$ _____

**4**   $4x + 31 =$ _____

You know how to graph a number.

**Example:** Graph 5.

**TRY IT!**

**5**   Graph 7.

**What You Will Learn**

*Lesson 3-1*

To solve an equation, isolate the variable by using inverse operations to "undo" the operations.

Find the value of $x$ in the equation $x - 2 = 6$.

$x - 2 = \phantom{+}6$    Given equation.
$\underline{+2 = +2}$    Subtract.
$x \phantom{- 2} = \phantom{+}8$    Simplify.

Find the value of $x$ in the equation $x \div 3 = 8$.

$x \div 3 = \phantom{2}8$    Given equation.
$\underline{\phantom{x \div 3}\cdot 3 \phantom{=} \cdot 3}$    Multiply.
$x \phantom{\div 3} = 24$    Simplify.

*Lesson 3-4*

The solutions of an inequality are a set of numbers. If the inequality symbol is $<$ or $>$, use an open circle on the graph. If the inequality symbol is $\leq$ or $\geq$, use a closed circle on the graph.

Graph $x \geq 5$.

All numbers greater than or equal to 5.

# Solve One-Step Equations

## KEY Concept

To solve an **equation**, isolate the variable by using **inverse operations** to "undo" the operations in equations.

### Addition and Subtraction Properties of Equality

$$x - 2 = 4$$      Given equation.

$$\underline{+2 \quad +2}$$      Add 2 to each side of the equation.

$$x = 6$$      Simplify.

### Multiplication and Division Properties of Equality

$$\frac{2x}{2} = \frac{108}{2}$$      Given equation.

         Divide each side by 2.

$$x = 54$$      Simplify.

Use an equation mat to model solving equations.

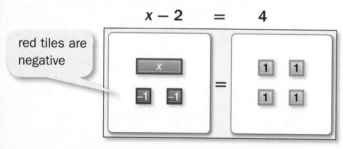

red tiles are negative

Add two positive tiles to both sides. Remove zero pairs.

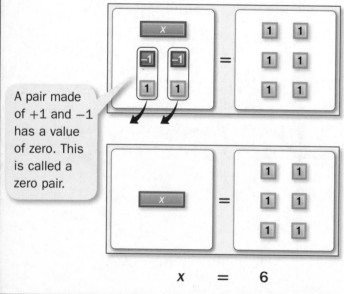

A pair made of +1 and −1 has a value of zero. This is called a zero pair.

$$x = 6$$

Check your solution to an equation by substituting your answer for the variable.

Copyright © Glencoe/McGraw-Hill, a division of The McGraw-Hill Companies, Inc.

### VOCABULARY

**Addition Property of Equality**
if you add the same number to each side of an equation, the two sides remain equal

**Division Property of Equality**
if you divide each side of an equation by the same non zero number, the two sides remain equal

**equation**
a mathematical sentence that contains an equal sign, =

**inverse operations**
operations that undo each other

**Multiplication Property of Equality**
if you multiply each side of an equation by the same number, the two sides remain equal

**Subtraction Property of Equality**
if you subtract the same number from each side of an equation, the two sides remain equal

## Example 1

**Solve 2x = −6 using algebra tiles.**

1. Model the equation.

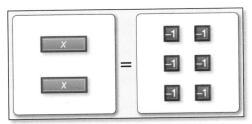

2. There are two *x*-tiles. Divide the tiles into 2 equal groups.

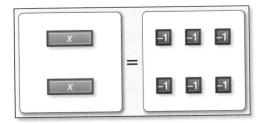

3. There are no zero pairs. There are three negative tiles for each variable tile. Write the solution.

*x* = −3

## YOUR TURN!

**Solve y + 4 = 6 using algebra tiles.**

1. Model the equation.

2. Make zero pairs. Add _____ negative tiles to each side.

3. Remove zero pairs and write the solution.

*y* = _____

## Example 2

**Solve 1.5 + r = 3. Check the solution.**

1. Use the inverse operation to solve.

$$1.5 + r = 3$$     Given equation.
$$\underline{-\,1.5 \qquad -1.5}$$     Subtract 1.5.
$$r = 1.5$$     Simplify.

2. Check the solution.

$$1.5 + 1.5 = 3 \checkmark$$

## YOUR TURN!

**Solve $\frac{2}{3}x = 8$. Check the solution.**

1. Use the inverse operation to solve.

$$\frac{2}{3}x = 8$$     Given equation.

$$\underline{\quad} \cdot \frac{2}{3}x = \underline{\quad} \cdot 8$$

$$x = \underline{\quad}$$     Simplify.

2. Check the solution.

$$\frac{2}{3}(\underline{\quad}) = 8$$

**GO ON**

# ▶ Guided Practice

**Solve using algebra tiles.**

**1** $z + 5 = -2$
Model the equation. Make zero pairs.

Remove zero pairs. Write the solution.

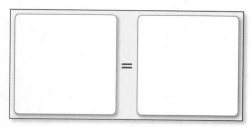

$z = \underline{\hspace{1.5cm}}$

**2** $5x = 25$
Model the equation. Group the tiles.

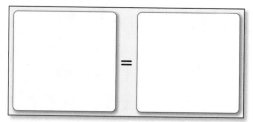

Write the solution.

$x = \underline{\hspace{1.5cm}}$

---

## Step by Step Practice

**3** Solve $-2.25 + y = -4$. Check the solution.

**Step 1** Use the inverse operation to solve.

$$-2.25 + y = -4 \qquad \text{Given equation.}$$

$$\underline{\hspace{5cm}} \qquad \underline{\hspace{4cm}}$$

$$y = \underline{\hspace{1.5cm}} \qquad \underline{\hspace{4cm}}$$

**Step 2** Check the solution.

$$-2.25 + y = -4$$

$$-2.25 + \underline{\hspace{2cm}} = -4$$

**Solve each equation. Check the solution.**

**4** $\frac{3}{4}x = 15$

_____ $\frac{3}{4}x =$ _____ $\cdot 15$

$x =$ _____

Check.

$\frac{3}{4}$ _____ $= 15$

**5** $x - 3 = 8$

_____

$x =$ _____

Check.

_____ $- 3 = 8$

**6** $x + 7 = -5$

_____

$x =$ _____

Check.

_____ $+ 7 = -5$

**7** $2t = -6$

_____ $=$ _____

$t =$ _____

Check.

_____ $=$ _____

## Step by Step Problem-Solving Practice

**Solve.**

**8** SHOPPING   The game Ivan wants to buy costs $42. He has already saved $19. How much more does he need to save?

Let $M$ equal the amount of money Ivan needs to save.

$M +$ _____ $=$ _____

_____

_____ $=$ _____

Check off each step.

_____ **Understand: I underlined key words.**

_____ **Plan: To solve the problem, I will** _____.

_____ **Solve: The answer is** _____.

_____ **Check: I checked my answer by** _____.

GO ON

## ▶ Skills, Concepts, and Problem Solving

**Solve each equation. Check the solution.**

**9** $y - 7 = 12$ _____

**10** $1.8 + z = -5$ _____

**11** $r + 4.5 = 10$ _____

**12** $-3 + b = -11$ _____

**13** $\frac{3}{5}x = 9$ _____

**14** $\frac{5}{4}y = -10$ _____

**15** $4 - h = -16$ _____

**16** $\frac{3}{7}m = 21$ _____

**Solve.**

**17** **TESTING**  To pass a test, Pat needs to get $\frac{2}{3}$ of the questions right. Pat knows she needs to get 42 questions right. How many questions are on the test?

_____

**18** **AREA**  Find the height of the triangle $h$ given that the area $A$ of the triangle is 30 cm² and the length of the base $b$ is 12 cm.

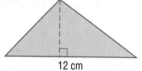

12 cm

$$A = \frac{1}{2}bh$$

_____

**Vocabulary Check**  **Write the vocabulary word that completes each sentence.**

**19** Operations that "undo" each other are called _____.

**20** The _____ states that if you add the same number to each side of an equation the two sides remain equal.

**21** A(n) _____ is a mathematical sentence that contains an equal sign.

**22** **Reflect**  Explain how a balanced scale is similar to a mathematical equation.

_____

_____

_____

🛑 STOP

# Solve Multi-Step Equations

## KEY Concept

You can model **equations** with algebra tiles.

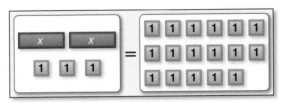

2x + 3  =  17

Add 3 negative tiles to each side and remove **zero pairs**.

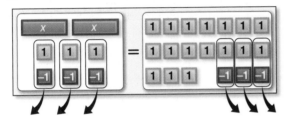

There are 2 *x*-tiles. Arrange the unit tiles into 2 equal groups.

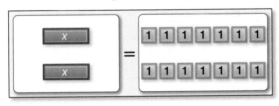

x = 7

Without a model, follow the **order of operations** in reverse to solve a multi-step equation. Use **inverse operations** to "undo" the operations in equations.

| | |
|---|---|
| 2x + 3 = 17 | Given equation. |
| − 3  − 3 | Subtract. |
| $\dfrac{2x}{2} = \dfrac{14}{2}$ | Simplify.  Divide. |
| x = 7 | Simplify. |

Check the solution to an equation by substituting the answer into the original equation for the variable.

### VOCABULARY

**equation**
a mathematical sentence that contains an equal sign, =

**inverse operations**
operations that undo each other

**order of operations**
rules that tell which operation to perform first when more than one operation is used

**zero pair**
a pair made of +1 and −1 algebra tiles with a value of zero

GO ON

## Example 1

**Solve 2x − 5 = 3 using algebra tiles.**

1. Model the equation.

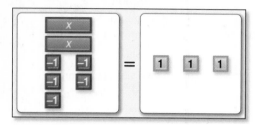

2. Make zero pairs. Add 5 positive tiles to each side.

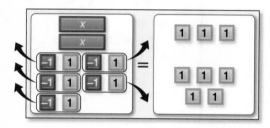

3. Remove zero pairs. There are 2 x-tiles. Divide the tiles into 2 equal groups.

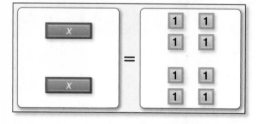

4. Write the solution.

   x = 4

**YOUR TURN!**

**Solve 3p + 2 = 8 using algebra tiles.**

1. Model the equation.

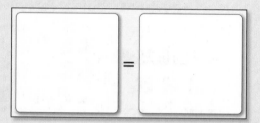

2. Make zero pairs. Add _____ tiles to each side.

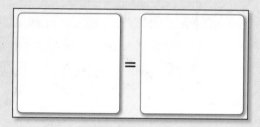

3. Remove zero pairs. There are ____ x-tiles. Divide the tiles into ____ equal groups.

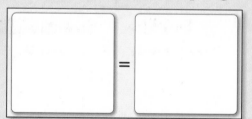

4. Write the solution.

   x = ____

## Example 2

**Solve $\frac{1}{10}n - 2.5 = 7.5$. Check the solution.**

1. Use inverse operations to solve.

| | |
|---|---|
| $\frac{1}{10}n - 2.5 = 7.5$ | Given equation. |
| $\underline{+2.5 \ +2.5}$ | Add 2.5. |
| $\frac{1}{10}n = 10$ | Simplify. |
| $10 \cdot \frac{1}{10}n = 10 \cdot 10$ | Multiply by 10. |
| $n = 100$ | Simplify. |

2. Check the solution.

$\frac{1}{10}(100) - 2.5 = 7.5$ ✔

---

**YOUR TURN!**

**Solve $50 - 3b = 35$. Check the solution.**

1. Use inverse operations to solve.

| | |
|---|---|
| $50 - 3b = 35$ | Given equation. |
| _____ | _____ |
| $-3b = $ _____ | Simplify. |
| _____ | _____ |
| $b = $ ____ | Simplify. |

2. Check the solution.

$50 - 3(\underline{\hspace{0.7cm}}) = \underline{\hspace{0.7cm}}$

---

 **Guided Practice**

**Solve each equation using algebra tiles.**

**1** $3b - 7 = -1$      $b = $ _____

**2** $2m + 5 = -3$      $m = $ _____

---

## Step by Step Practice

**3** Solve $1.5r + 3 = 10$. Label each step. Check the solution.

**Step 1** Use inverse operations to solve.

$1.5r + 3 = 9$      _____

_____      _____

_____ = _____      _____

_____      _____

$r = $ _____      _____

**Step 2** Check the solution. $1.5(\underline{\hspace{0.7cm}}) + 3 = 9$

**GO ON**

**Solve each equation. Label each step. Check each solution.**

**4** $6y + 9 = -15$      Given equation.

_____      Subtract.

_____ = _____      Divide.

$y =$ _____      Simplify.

Check.

$6(\_\_\_\_\_) + 9 = -15$

**5** $\dfrac{2}{3}x - 5.5 = 12.5$      Given equation.

_____    _____

_____ = _____    _____

_____ = _____    _____

$x =$ _____    _____

Check.

$\dfrac{2}{3}(\_\_\_\_\_) - 5.5 = 12.5$

## Step by Step Problem-Solving Practice

**Solve.**

**6** **WORK** The equation $\$82 = 7h + 19$ represents how much money Randy earned last week. The number of hours he worked is represented by $h$. How many hours did Randy work last week?

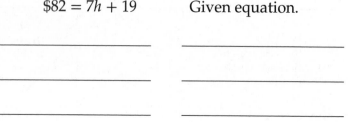        $\$82 = 7h + 19$      Given equation.

_____    _____

_____    _____

_____    _____

_____    _____

Check off each step.

_____ **Understand: I underlined key words.**

_____ **Plan: To solve the problem, I will** _____.

_____ **Solve: The answer is** _____.

_____ **Check: I checked my answer by** _____.

## ▶ Skills, Concepts, and Problem Solving

**Solve each equation. Check each solution.**

**7** $-6x + 12 = 30$

**8** $41 - 5y = 16$

**9** $\frac{1}{3}b + 11 = 19$

**10** $\frac{3}{4}z - 5 = 7$

**11** $-2x - 23 = -55$

**12** $\frac{1}{8}t + 6 = -10$

**Solve.**

**13** PARTY PLANNING  The cost for a company to cater a dinner party is represented by the equation $P = 8.5n + 150$, where $P$ is the cost and $n$ is the number of people. If the dinner party cost $269, how many people are at the party?

_____

**14** ANGLES  Find the value of $x$ in the triangle at the right. (Hint: There are 180 degrees in a triangle.)

_____

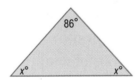

**Vocabulary Check**  **Write the vocabulary word that completes each sentence.**

**15** A pair made of +1 and −1 algebra tiles is called a(n) _____.

**16** The _____ are rules that tell you which operation to perform first when more than one operation is used.

**17** Reflect  Explain this statement: "Every time you solve an equation, you should know if your answer is correct."

_____

_____

STOP

**Solve each equation using algebra tiles.**

**1** $2x - 4 = 6$

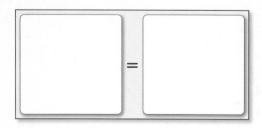

$x =$ _____

**2** $3x - 1 = -10$

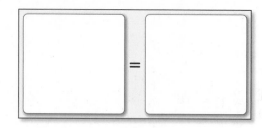

$x =$ _____

**Solve each equation. Check each solution.**

**3** $x - 1 = -6$

**4** $x + 12 = 30$

**5** $10x = 30$

**6** $2x - 3 = 3$

**7** $3x + 7 = -11$

**8** $-2x + 6 = 44$

**Solve.**

**9** **FUNDRAISING** For a fundraiser, the prom committee sold 324 rolls of wrapping paper. After paying $75 for the supplies, the committee had $1,059. What was the sale price of each roll of wrapping paper?

_____

**10** **GRADES** On his final report card, Jerry received 14 A's. This is 3 more than his sister Jessica. How many A's did Jessica receive?

_____

**11** **COLLECTING** Sierra bought a doll 5 years ago. According to a collector magazine, the doll is now worth 4 more than 3 times the amount she paid for it. The doll is selling for $142. How much did Sierra originally pay for the doll?

_____

# Solve Equations with Variables on Both Sides

## KEY Concept

To solve an **equation** with variables on both sides, first use inverse operations to isolate the variable and solve the equation.

Choose the first **inverse operation** so that the variable term stays positive, if possible.

| Positive Result | Negative Result |
|---|---|
| $4x + 2 = 5x - 1$ | $4x + 2 = 5x - 1$ |
| $\underline{-4x \qquad -4x}$ | $\underline{-5x \quad -5x}$ |
| $2 = x - 1$ | $-x + 2 = -1$ |

| | |
|---|---|
| $4x + 2 = 5x - 1$ | Given equation. |
| $\underline{-4x \qquad -4x}$ | Subtraction Property |
| $2 = x - 1$ | Simplify. |
| $\underline{+1 \qquad +1}$ | Addition Property |
| $3 = x$ | Simplify. |

VOCABULARY

**equation**
a mathematical sentence that contains an equal sign, =

**inverse operations**
operations that undo each other

You can use algebra tiles to model these types of equations as well.

### Example 1

**Solve $5p = 4p + 7$.**

1. Find a positive variable. Use inverse operations.

$$5p = 4p + 7$$
$$\underline{-4p \quad -4p}$$
$$1p = 7$$

2. Write the solution.

$$p = 7$$

### YOUR TURN!

**Solve $-5 + 2q = 3q$.**

1. Find a positive variable. Use inverse operations.

$$-5 + 2q = 3q$$
$$\underline{\qquad\qquad}$$
$$\underline{\qquad} = \underline{\qquad}$$

2. Write the solution.

$$q = \underline{\qquad}$$

GO ON

## Example 2

Solve $4h - 2 = 2h + 6$.

1. Find a positive variable.
   Use inverse operations.

$$4h - 2 = 2h + 6$$
$$\underline{-2h \qquad -2h}$$
$$2h - 2 = 6$$

2. Isolate the variable.

$$2h - 2 = 6$$
$$\underline{+2 \quad +2}$$
$$2h = 8$$

3. Simplify.

$$\frac{2h}{2} = \frac{8}{2}$$

4. Write the solution.        $h = 4$

### YOUR TURN!

Solve $-12 + 6a = 9a - 3$.

1. Find a positive variable.
   Use inverse operations.

$$-12 + 6a = 9a - 3$$
$$\underline{\hspace{5cm}}$$
$$\underline{\hspace{2cm}} = \underline{\hspace{2cm}}$$

2. Isolate the variable.   $\underline{\hspace{2cm}} = \underline{\hspace{2cm}}$

$$\underline{\hspace{5cm}}$$
$$\underline{\hspace{2cm}} = \underline{\hspace{2cm}}$$

3. Simplify.

4. Write the solution.        $a = \underline{\hspace{1.5cm}}$

---

## ▶ Guided Practice

**Solve each equation.**

**1**   $8y = 7y + 21$

$$\underline{\hspace{3cm}}$$
$$\underline{\hspace{1.5cm}} = \underline{\hspace{1.5cm}}$$

**2**   $4m - 7 = 5m$

$$\underline{\hspace{2.5cm}}$$
$$\underline{\hspace{1.5cm}} = \underline{\hspace{1.5cm}}$$

---

## Step by Step Practice

**3**   Solve $9g - 10 = 5g + 6$. Label each step.   $9g - 10 = 5g + 6$

**Step 1**   Find a positive variable. Use inverse operations.        $\underline{\hspace{3cm}}$

**Step 2**   Isolate the variable.        $\underline{\hspace{2cm}} = \underline{\hspace{1cm}}$

**Step 3**   Simplify.        $\underline{\hspace{3cm}}$

**Step 4**   Write the solution.   $g = \underline{\hspace{1.5cm}}$

$g = \underline{\hspace{1.5cm}}$

**Solve each equation. Label each step.**

**4**  $5x + 9 = 7x - 1$

| | |
|---|---|
| _____ | Subtract. |
| $9 = $ _____ $- 1$ | Simplify. |
| _____ | Add. |
| _____ $= $ _____ | Simplify. |
| _____ | Divide. |
| _____ $= $ _____ | Simplify. |

**5**  $10z - 14 = 7 + 3z$

| | |
|---|---|
| _____ | _____ |
| _____ $=$ _____ | _____ |
| _____ | _____ |
| _____ $=$ _____ | _____ |
| _____ | _____ |
| _____ $=$ _____ | _____ |

## Step by Step Problem-Solving Practice

**Solve.**

**6**  **NUMBER SENSE**   Six plus twelve times a number is 14 more than eight times the number. Find the number.

$6 + 12x = 14 + 8x$

| | |
|---|---|
| _____   _____ | _____ |
| _____ $=$ _____ | _____ |
| _____ | _____ |
| _____ $=$ _____ | _____ |
| _____ | _____ |
| _____ $=$ _____ | _____ |

Check off each step.

_____ Understand: I underlined key words.

_____ Plan: To solve the problem, I will _____

_____.

_____ Solve: The answer is _____.

_____ Check: I checked my answer by _____.

**GO ON**

**Skills, Concepts, and Problem Solving**

**Solve each equation. Check the solution.**

**7** $-3x = -4x + 7$

**8** $4y = 3y + 17$

**9** $12y - 9 = 18 + 3y$

**10** $8t - 3 = 2t - 15$

**11** $-21x - 24 = 30 - 3x$

**12** $5z - 1 = -3z + 4$

**Solve.**

**13** **NUMBER SENSE** Thirteen less than twice a number is 2 more than half the number. Find the number.

_____

**14** **CAR RENTAL** Kathleen is trying to rent a car for a trip. She is comparing rental car companies using the chart below. How many miles would she have to drive if the price was going to be the same for both Company ABC and Company XYZ?

| Company ABC Costs | Company XYZ Costs |
|---|---|
| $15 a day | $21 a day |
| $0.25 per mile | $0.13 per mile |

_____

**Vocabulary Check** **Write the vocabulary word that completes each sentence.**

**15** Multiplication and division are _____ because they undo each other.

**16** When solving a(n) _____, use inverse operations to undo the operations.

**17** **Reflect** Write a sentence that can be translated into an equation but not in the same order as the words. What must be true about your sentence? Explain your answer.

_____

_____

**STOP**

# Solve One-Step Inequalities

## KEY Concept

### Addition and Subtraction Properties of Inequality

To simplify an **inequality**, use **inverse operations** to isolate the variable.

| | |
|---|---|
| $x - 32 > 14$ | Given equation. |
| $\underline{+32 \quad +32}$ | Add 32. |
| $x > 46$ | Simplify. |

### Multiplication and Division Properties of Inequality

Multiplying or dividing the same negative number from each side of an inequality reverses the inequality symbol and results in a true inequality.

| | |
|---|---|
| $\dfrac{-2x}{-2} \geq \dfrac{108}{-2}$ | Given equation. Divide by −2. |
| $x \leq -54$ | Simplify. |

> × or ÷ by (−) flips the sign

### Graphing Solutions

The solutions of an inequality are a set of numbers instead of a single value. If the inequality symbol is $<$ or $>$, use an open circle on the graph. If the inequality symbol is $\leq$ or $\geq$, use a closed circle on the graph.

| Symbols | Words | Graph |
|---|---|---|
| $x > 46$ | all numbers greater than 46 | 45 46 47 48 49 |
| $x < 46$ | all numbers less than 46 | 42 43 44 45 46 |
| $x \leq -54$ | −54 and all numbers less than −54 | −57 −56 −55 −54 −53 |
| $x \geq -54$ | −54 and all numbers greater than −54 | −55 −54 −53 −52 −51 |

Check the solution by substituting the value back into the inequality. Check the graph by substituting values on both sides of the number line around the circle.

## VOCABULARY

**Addition Property of Inequality**
if you add the same number to each side of an inequality, the inequality remains true

**Division Property of Inequality**
if you divide each side of an inequality by the same positive number the inequality remains true: If you divide each side of an inequality by a negative number and use the inverse inequality symbol, the inequality remains true

**inequality**
an open sentence that contains the symbol $<$, $\leq$, $>$, or $\geq$

**inverse operations**
operations that undo each other

**Multiplication Property of Inequality**
if you multiply each side of an inequality by the same nonnegative number the inequality remains true: If you multiply each side of an inequality by a negative number and use the inverse inequality symbol, the inequality remains true

**Subtraction Property of Inequality**
if you subtract the same number from each side of an inequality, the inequality remains true

GO ON

## Example 1

**Solve y + 7 < 12. Graph the solution on a number line.**

1. Use inverse operations to solve.

| | |
|---|---|
| $y + 7 < 12$ | Given equation. |
| $\underline{-7 \quad -7}$ | Subtract 7. |
| $y < 5$ | Simplify. |

2. Graph the solution. Place an open circle on 5 and draw the arrow left.

3. Check the solution and graph.

   **4** is left of 5, $4 + 7 < 12$ ✔

   **6** is right of 5, $6 + 7 \not< 12$

## YOUR TURN!

**Solve m − 15 ≤ 3. Graph the solution on a number line.**

1. Use inverse operations to solve.

   $m - 15 \leq 3$ _____

   _____ _____

   $m \leq$ _____ _____

2. Graph the solution. Place _____

   on _____ and draw the arrow _____.

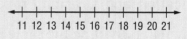

3. Check the solution and graph.

   _____ is left of 18, _____ − 15 ◯ 3

   _____ is right of 18, _____ − 15 ◯ 3

## Example 2

**Solve −6w ≤ 18. Graph the solution on a number line.**

1. Use inverse operations to solve.

| | |
|---|---|
| $-6w \leq 18$ | Given equation. |
| $\underline{\div(6) \quad \div(6)}$ | Divide by −6. |
| $w \geq -3$ | Simplify. |

> × or ÷ by (−) flips the sign

2. Graph the solution.
   Place a closed circle on −3 and draw the arrow right.

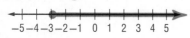

## YOUR TURN!

**Solve $\frac{d}{3} > 4$. Graph the solution on a number line.**

1. Use inverse operations to solve.

   $\frac{d}{3} > 4$ _____

   _____ $\frac{d}{3} > 4$ _____ _____

   $d$ ◯ _____ _____

2. Graph the solution.

   Place _____ on _____ and

   draw the arrow _____.

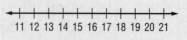

**Guided Practice**

**Solve the inequality. Graph the solution on a number line.**

**1** $x + 5 > 1$ Given equation.

_____ _____

$x >$ _____ _____

Graph the solution. Place _____ on _____

and draw the arrow _____.

Check the solution and graph.

_____ is left of $-4$, _____ $+ 5 \bigcirc 1$

_____ is right of $-4$, _____ $+ 5 \bigcirc 1$

**Step by Step Practice**

**2** Solve $-\dfrac{1}{2} z \le -4$. Label each step. Graph the solution on a number line.

**Step 1** Use inverse operations.

$-\dfrac{1}{2} z \le -4$ Given equation.

$(-2) -\dfrac{1}{2} z \le -4(-2)$ _____

$z \bigcirc$ _____ _____

**Step 2** Graph the solution.

**Step 3** Check the solution and graph.

_____ is left of _____, $-\dfrac{1}{2} ($ _____ $) \bigcirc -4$

_____ is right of _____, $-\dfrac{1}{2} ($ _____ $) \bigcirc -4$

GO ON

**Solve each inequality. Graph the solution on a number line.**

**3**  $7 + s < 12$

_____

$s \bigcirc$ _____

0  1  2  3  4  5  6  7  8  9  10

**4**  $-3 + b \geq 2$

_____

$b \bigcirc$ _____

0  1  2  3  4  5  6  7  8  9  10

**5**  $\dfrac{x}{5} > 2$

_____

$x \bigcirc$ _____

5  6  7  8  9  10 11 12 13 14 15

**6**  $-2c \leq 8$

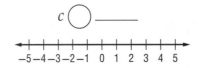

Remember, × or ÷ by (−) flips the inequality symbol.

_____

$c \bigcirc$ _____

−5 −4 −3 −2 −1  0  1  2  3  4  5

## Step by Step Problem-Solving Practice

**Solve.**

**7**  Roberta is more than twice as old as Jay. If Roberta is 48 years old, what are the possibilities for Jay's age?

$2j < 48$        Given equation.

_____        _____

$j \bigcirc$ _____        _____

Check off each step.

_____ **Understand: I underlined key words.**

_____ **Plan: To solve the problem, I will** _____.

_____ **Solve: The answer is** _____.

_____ **Check: I checked my answer by** _____

_____.

**90  Chapter 3** Equations and Inequalities

## Skills, Concepts, and Problem Solving

**Solve each inequality. Graph each solution on a number line.**

**8** $11 > y + 3$ ________

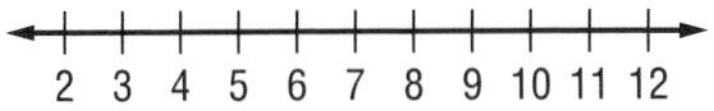


**9** $n + 23 > 25$ ________

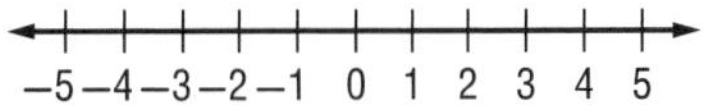


**10** $-6 + r \leq -15$ ________

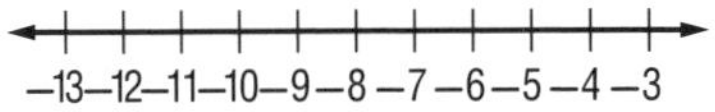


**11** $12z < -84$ ________

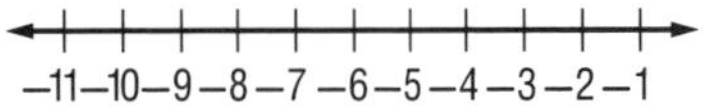


**12** $9p \geq -180$ ________

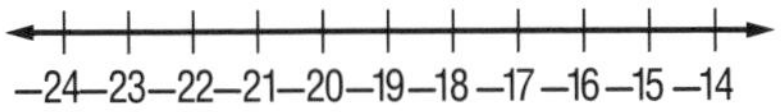


**13** $-4h \geq 20$ ________

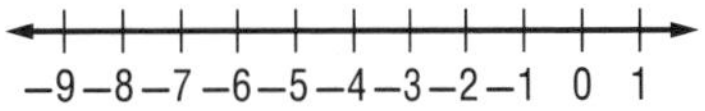


**Solve.**

**14** **AREA** The area of the rectangle is greater than 50 square inches. How long does the length need to be?

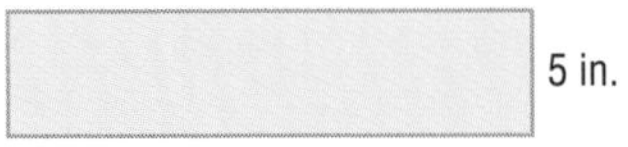


________________________

**15** **SALES** Gary is selling magazines for a fundraiser. He earns \$2.50 for each magazine he sells. If Gary wants to earn more than \$30, how many magazines does he need to sell?

________________________

**Vocabulary Check** **Write the vocabulary word that completes each sentence.**

**16** The ________________________ states that if you subtract the same number from each side of an inequality, the inequality remains true.

**17** A number sentence that compares two unequal expressions is a(n) ________________________.

**18** **Reflect** Explain the difference between the solution of an equation and the solution of an inequality.

________________________

________________________

________________________

STOP

**Solve each equation.**

**1** $12 + 1.5r = 3r$

**2** $6a = 26 + 4a$

**3** $15 - t = 23 - 2t$

**4** $7x + 2 = 3x + 94$

**Solve each inequality. Graph each solution on a number line.**

**5** $y + 7 > 21$

**6** $23 + s \leq 14$

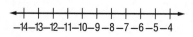

**7** $x + 5 \geq -3$

**8** $3t \geq 21$

**9** $-4 \leq \dfrac{g}{-2}$

**10** $-3u < -21$

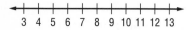

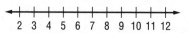

**Solve.**

**11** **HOME IMPROVEMENT**   A remodeling company charges $32 per hour plus $125 per day. Another company charges $25 per hour plus $167 per day. After how many hours in one day would the companies charge the same rate?

_____

**12** **SAVINGS**   Cherie is saving money from babysitting. She wants to save at least $350. She gets paid $7 per hour of babysitting. How many hours must she work to save up the money?

_____

# Solve Multi-Step Inequalities

<table>
<tr>
<td>

## KEY Concept

Solving **inequalities** differs from solving equations in three ways.

- An inequality symbol is used in place of an equal sign.

- You change the direction of the inequality symbol when you multiply or divide by a negative number.

- The solution is a range of numbers shown on a number line.

When graphing inequalities the circle and shading depends on the symbol.

| | |
|---|---|
| open circle: $<$ and $>$ | closed circle: $\leq$ and $\geq$ |
| shade left: $<$ and $\leq$ | shade right: $>$ and $\geq$ |

</td>
<td>

## VOCABULARY

**inequality**
an open sentence that contains the symbol $<$, $\leq$, $>$, $\geq$

**order of operations**
rules that tell which operation to perform first when more than one operation is used

</td>
</tr>
</table>

When solving a multi-step inequality, follow the order of operations in reverse.

## Example 1

Solve $\dfrac{f}{3} + 18 > 24$. Graph the solution on a number line.

1. Use inverse operations to solve.

$\dfrac{f}{3} + 18 > 24$    Given equation.

$\underline{-18 \quad -18}$    Subtract 18.

$\dfrac{f}{3} > 6$    Simplify.

$\dfrac{f}{3} \cdot 3 > 6 \cdot 3$    Multiply by 3.

$f > 18$    Simplify.

2. Graph the solution. Place an open circle on 18 and draw the arrow right.

## YOUR TURN!

Solve $-2x + 3 \geq 9$. Graph the solution on a number line.

1. Use inverse operations to solve.

$-2x + 3 \geq 9$    Given equation.

_____    _____.

____ $\geq$ ____    _____.

_____    _____.

2. Graph the solution. Place

_____ circle on ____ and

draw the arrow ____.

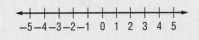

**GO ON**

## Example 2

Solve $2d + 4 < 6 + d$. Graph the solution on a number line.

1. Use inverse operations to solve.

$$2d + 4 < 6 + d$$

$$\underline{\phantom{xx}-d \qquad\qquad -d\phantom{xx}} \quad \text{Subtract } d.$$

$$d + 4 < 6 \qquad\quad \text{Simplify.}$$

$$\underline{\phantom{xx}-4 \;\; -4\phantom{xx}} \qquad\quad \text{Subtract 4.}$$

$$d < 2 \qquad\qquad \text{Simplify.}$$

2. Graph the solution. Place an open circle on 2 and draw an arrow left.

## YOUR TURN!

Solve $6w + 3 \geq -w + 2$. Graph the solution on a number line.

1. Use inverse operations to solve.

$$6w + 3 \geq -w + 2 \qquad \text{Given equation.}$$

_____          _____

_____ $\geq$ _____          _____

_____          _____

_____ $\geq$ _____          _____

_____          _____

_____ $\geq$ _____          _____

2. Graph the solution. Place _____ circle on _____ and draw an arrow _____.

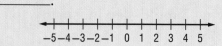

---

**Guided Practice**

Solve each inequality. Graph the solution on a number line.

1. $\dfrac{x}{5} - 12 > 7$

$$\dfrac{x}{5} - 12 > 7 \qquad\qquad \text{Given equation.}$$

_____          Add.

_____ > _____          Simplify.

_____ > _____          Multiply.

_____ > _____          Simplify.

Place _____ circle on _____ and draw an arrow _____.

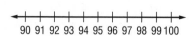

**Solve the inequality. Graph the solution on a number line.**

**2**   $4 + 6y > 40$

$4 + 6y > 40$         Given equation.

_____        _____

___ > ___        _____

_____        _____

___ ◯ ___        _____

Place _____ circle on _____ and

draw an arrow _____.

```
<--+--+--+--+--+--+--+--+--+--+--+-->
   0  1  2  3  4  5  6  7  8  9 10
```

---

**Step by Step Practice**

**3**   Solve $-8 - 10t < -12t - 21$. Graph the solution on a
number line.

$-8 - 10t <$ _____        Given equation.

_____       _____

___ < _____      _____

_____       _____

___ < _____      _____

_____       _____

___ ◯ _____      _____

Place _____ circle on _____ and draw an arrow _____.

```
<--+--+--+--+--+--+--+--+--+--+--+-->
 -12-11-10-9 -8 -7 -6 -5 -4 -3 -2
```

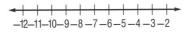

GO ON

**Solve each inequality. Label each step. Graph each solution on a number line.**

**4**  $48 - 20n \le 7n - 6$

_____ $-$ _____ $\le$ _____ $- 6$        Given equation.

_____        _____

_____        _____

_____        _____

_____        _____

_____        _____

_____        _____

Place _____ circle on _____

and draw an arrow _____.

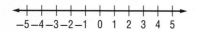

**5**  $12 - 4m \le -7m - 3$

_____        _____

_____        _____

_____        _____

_____        _____

_____        _____

_____        _____

Place _____ circle on _____

and draw an arrow _____.

## Step by Step Problem-Solving Practice

**Solve.**

**6** TEST SCORES  Lance knows the inequality $6x - 18 \geq 72$ shows the quiz score he must earn to get an A in math class. What is the minimum score Lance can earn on his quiz?

$$6x - 18 \geq 72 \qquad \text{Given equation.}$$

_____   _____

_____   _____

_____   _____

Check off each step.

_____ **Understand: I underlined key words.**

_____ **Plan: To solve the problem, I will** _____.

_____ **Solve: The answer is** _____.

_____ **Check: I checked my answer by** _____.

## Skills, Concepts, and Problem Solving

**Solve each inequality. Graph each solution on a number line.**

**7**  $-4z + 11 < 23$   _____

$$\xleftarrow{\phantom{xx}}\!\!+\!\!+\!\!+\!\!+\!\!+\!\!+\!\!+\!\!+\!\!+\!\!+\!\!+\!\!\xrightarrow{\phantom{xx}}$$
$$-5\,-4\,-3\,-2\,-1\ \ 0\ \ 1\ \ 2\ \ 3\ \ 4\ \ 5$$

**8**  $5r - 20 \leq 45$   _____

$$\xleftarrow{\phantom{xx}}\!\!+\!\!+\!\!+\!\!+\!\!+\!\!+\!\!+\!\!+\!\!+\!\!+\!\!+\!\!\xrightarrow{\phantom{xx}}$$
$$8\ \ 9\ \ 10\ 11\ 12\ 13\ 14\ 15\ 16\ 17\ 18$$

**9**  $5x + 10 > 3x + 14$   _____

$$\xleftarrow{\phantom{xx}}\!\!+\!\!+\!\!+\!\!+\!\!+\!\!+\!\!\xrightarrow{\phantom{xx}}$$
$$-2\ \ -1\ \ 0\ \ 1\ \ 2\ \ 3\ \ 4$$

**10**  $-6k - 1 \leq 3k + 8$   _____

$$\xleftarrow{\phantom{xx}}\!\!+\!\!+\!\!+\!\!+\!\!+\!\!+\!\!+\!\!\xrightarrow{\phantom{xx}}$$
$$-4\ \ -3\ \ -2\ \ -1\ \ 0\ \ 1\ \ 2\ \ 3$$

GO ON

**Solve each inequality. Graph each solution on a number line.**

**11** $42 - 3x \geq 24x + 60$ _____

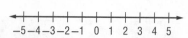

**12** $24 + 16y > 30 + 13y - 21$ _____

**13** $2y + 5 \leq 8 + y - 7$ _____

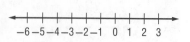

**14** $12t - 6 - 8t > -6t + 4$ _____

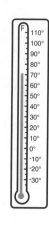

**Solve.**

**15** **TEMPERATURE**   The weather forecaster said that 15°F more than double the low temperature is greater than the high temperature for the day. If the high temperature is 71°F, what is the lowest temperature for the day?

_____

**16** **INTEGERS**   The sum of two consecutive integers is greater than 49. What is the minimum value for the smallest integer?

_____

**Vocabulary Check**   **Write the vocabulary word that completes each sentence.**

**17** A(n) _____ is an open sentence that contains the symbol $<$, $>$, $\leq$, or $\geq$.

**18** The set of rules that tell you which operation to do first is called the

_____.

**19** **Reflect**   Compare the steps used to solve an equation and the steps used to solve an inequality.

_____

_____

_____

_____

**STOP**

# Solve for a Specific Variable

## KEY Concept

To solve an **equation** for a specific variable, first identify the operations that are being performed on that variable. Then undo them by using inverse operations.

The **formula** for the circumference of a circle is $C = 2\pi r$.

The circumference is the distance around the circle.

In the circumference formula the expression $2\pi r$ could have been written as $2 \cdot \pi \cdot r$. The terms of the expression are multiplied together. The inverse operation of multiplication is division. In order to solve for the variable $r$, use division.

$C = 2\pi r$    Write the formula.

$\dfrac{C}{2\pi} = \dfrac{2\pi r}{2\pi}$    Divide by $2\pi$.

$\dfrac{C}{2\pi} = \dfrac{\cancel{2\pi} r}{\cancel{2\pi}}$    Cancel like terms.

$\dfrac{C}{2\pi} = r$    Simplify.

According to the principles of equality, the solution can be also written as $r = \dfrac{C}{2\pi}$.

## VOCABULARY

**equation**
   a mathematical sentence that contains an equal sign, $=$

**formula**
   an equation that states a rule for the relationship between certain quantities

The formula for the circumference of a circle shows the relationship between the value of pi ($\pi$), the radius of a circle $r$, and the circumference of a circle $C$.

## Example 1

**Solve $f = g - 1$ for $g$.**

1. Locate the term with the variable $g$. $g$

2. Perform the inverse operations.

$$
\begin{array}{r}
f = g - 1 \\
\underline{+1 \quad\quad +1} \\
f + 1 = g
\end{array}
$$

## YOUR TURN!

**Solve $3a = b$ for $a$.**

1. Locate the term with the variable. _____

2. Perform the inverse operations.

$$3a = b$$

$$\underline{\quad\quad} = \underline{\quad\quad}$$

$$a = \underline{\quad\quad}$$

GO ON

## Example 2

Solve $\dfrac{3b}{5} = 9z$ for $b$.

1. Locate the term with the variable $b$. **3b**

2. Perform the inverse operations.

$$\dfrac{3b}{5} = 9z$$

$$\dfrac{3b}{5} \cdot 5 = 9z \cdot 5$$

$$3b = 45z$$

$$\dfrac{3b}{3} = \dfrac{45z}{3}$$

$$b = 15z$$

**YOUR TURN!**

Solve $y = mx + b$ for $x$.

1. Locate the term with the variable $x$. _____

2. Perform the inverse operations.

$$y = mx + b$$

_____

_____ = _____

_____ = _____

_____ = _____

 ## Guided Practice

**Solve each equation for the given variable.**

**1** $y = 5x$ for $x$

Locate the term with the variable $x$. _____

$$y = 5x$$

_____ = _____

_____ = _____

**2** $a - 10 = b$ for $a$

Locate the term with the variable $a$. _____

$$a - 10 = b$$

_____

_____ = _____

## Step by Step Practice

**3** The formula for the perimeter of a rectangle is $P = 2\ell + 2w$.
Solve the formula for $\ell$.

**Step 1** Locate the term with the variable $\ell$. _____

**Step 2** Perform the inverse operations.

$$P = 2\ell + 2w$$

_____

_____

_____

**Solve each equation for the given variable.**

**4** $I = prt$ for $r$

$$I = prt$$

_____

_____

**5** $ax + 13 = c$ for $x$

$$ax + 13 = c$$

_____

_____

_____

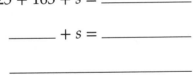

## Step by Step *Problem-Solving Practice*

**Solve.**

**6** **BOWLING**  Julia knows the following equation will show her bowling average for 3 games. *A* represents her average and *s* represents her score for her third game. Solve the equation for *s*.

$$\frac{125 + 163 + s}{3} = A$$

$$\frac{125 + 163 + s}{3} \rule{1cm}{0.4pt} = A \rule{2cm}{0.4pt}$$

$$125 + 163 + s = \rule{2cm}{0.4pt}$$

$$\rule{1.5cm}{0.4pt} + s = \rule{2cm}{0.4pt}$$

$$\rule{3cm}{0.4pt}$$

$$\rule{2cm}{0.4pt} = \rule{2cm}{0.4pt}$$

Check off each step.

_____ **Understand: I underlined key words.**

_____ **Plan: To solve the problem, I will** _____.

_____ **Solve: The answer is** _____.

_____ **Check: I checked my answer by** _____.

GO ON

## ▶ Skills, Concepts, and Problem Solving

**Solve each equation for the given variable.**

**7** $r + 2b = 1$ for $b$ _____

**8** $P = 4s - 9$ for $s$ _____

**9** $7g + 14 - 11 = f$ for $g$ _____

**10** $A = \frac{1}{2}bh$ for $h$ _____

**11** $\dfrac{15p + 22}{12} = 8m$ for $p$ _____

**12** $y = mx + b$ for $b$ _____

**Solve.**

**13** **RENTAL CAR**  The price of a rental car is given by the equation $C = 22 + 0.18m$, where $C$ is the total cost and $m$ is the miles driven. Solve the equation for $m$.

_____

**14** **TRAVEL**  Tara knows the equation $t = \dfrac{d}{r} + 1.5$ represents how long a trip will take her if she stops for an hour and a half lunch. Solve the equation for $d$.

_____

**15** **AREA**  Write an equation for the area of the pentagon if the area of the triangle is 24 cm². Then solve for $w$. Hint: The formula for the area of a rectangle is $A = \ell w$.

_____

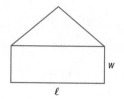

**Vocabulary Check**  **Write the vocabulary word that completes each sentence.**

**16** A(n) _____ is a mathematical sentence that contains an equal sign.

**17** An equation that states a rule for the relationship between certain quantities is called a(n) _____.

**18** **Reflect**  Why is it important to be able to solve an equation for a different variable? Explain your answer and give an example.

_____

_____

_____

**STOP**

**Solve each inequality. Graph each solution on a number line.**

**1** $3d - 2 \geq 19$

**2** $5y - 1 < 19$

**3** $5 - 2w < -11$

**4** $-3t + 7 \geq 10$

**5** $\dfrac{2x}{5} \geq -6$

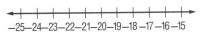

**6** $\dfrac{a + 6}{3} > -5$

**Solve each equation for the given variable.**

**7** $s = \dfrac{1}{2}m + k$ for $m$ _____

**8** $y = 2x + 7$ for $x$ _____

**9** $10 = r + 4d$ for $d$ _____

**10** $a = b - c + d$ for $c$ _____

**Solve.**

**11 PARTY PLANNING** It costs $2,500 to rent a gymnasium for a party. A committee received a donation of $700 and plans to charge $5.00 per person for admission. How many people must attend for the committee to make a profit? Write an inequality and solve.

_____

**12 MEDICINE** Clark's Rule for determining medicine doses for children is given by the formula $c = a \cdot \dfrac{w}{150}$, where $c$ is the child's dosage amount, $a$ is the adult dosage amount, and $w$ is the weight of the child in pounds. Solve the formula for $w$.

_____

# Chapter Test

**Solve each equation. Check the solution.**

**1** $d - 15 = 2$

**2** $34 - w = 6$

**3** $h + 15 = 22$

**4** $2k = 44$

**5** $\dfrac{w}{3} = 4$

**6** $-4f = 36$

**7** $2x + 3 = 23$

**8** $-4y - 2 = -14$

**9** $\dfrac{2c}{5} + 1 = 3$

**10** $\dfrac{z}{4} - 15 = 3$

**11** $15 - g = 23 - 2g$

**12** $4y - 6 = 2y + 8$

**Solve each inequality. Graph each solution on a number line.**

**13** $x - 3 > 4$ _____

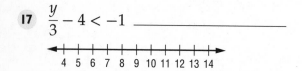

**14** $-3t \geq 24$ _____

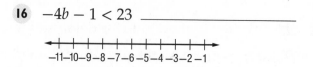

**15** $2f + 2 \geq 16$ _____

**16** $-4b - 1 < 23$ _____

**17** $\dfrac{y}{3} - 4 < -1$ _____

**18** $15 - 3n \geq 30$ _____

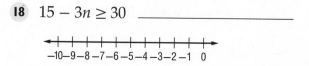

**Solve each equation for the given variable.**

**19** $ab - ac = 1$ for $b$

**20** $3g + 2 = f$ for $g$

**21** $\dfrac{h+1}{2} = rs$ for $h$

**22** $V = Bh$ for $h$

**23** $nx = 3$ for $x$

**24** $h + 6 = vt$ for $v$

**25** $K = 5h - \dfrac{1}{2}$ for $h$

**26** $x = \dfrac{-b}{2a}$ for $b$

**Solve. Write each answer in simplest form.**

**27** **GEOMETRY**  The area of a parallelogram is given by the formula $A = bh$. Find the height of the parallelogram with an area of 85 ft² and a base of 17 ft.

_____

**28** **SPORTS**  In order to play in the finals, the school basketball team must win at least 21 games. So far this season, they have won 13 games. What is the least amount of games they must win to play in the finals?

_____

**29** **MANUFACTURING**  The cost of manufacturing widgets is given by the equation $C = 2p + 35$. Find the price $p$ per widget if the cost of manufacturing $C$ is $65.

_____

**30** **TEMPERATURE**  The formula $F = \dfrac{9}{5}C + 32$ converts a temperature in degrees Celsius to degrees Fahrenheit. Solve the formula for $C$. What is 65°F in Celsius?

_____

STOP

# Linear Equations

## How much money can you earn?

The amount of money Emily can earn at a summer job is given by the linear equation $y = 7x$, where $y$ represents the total dollars earned and $x$ represents the number of hours worked. You can find solutions of this equation to find how much money Emily could make.

STEP **2** Preview    Get ready for Chapter 4. Review these skills and compare them with what you will learn in this chapter.

| What You Know | What You Will Learn |
|---|---|

**What You Know**

You know how to write a fraction in simplest form.

**Example:** $\dfrac{6}{12}$

$$\dfrac{6 \div 6}{12 \div 6} = \dfrac{1}{2}$$

**TRY IT!**

**1** $\dfrac{7}{21} = \dfrac{\square}{\square}$    **2** $\dfrac{15}{20} = \dfrac{\square}{\square}$

**3** $\dfrac{44}{66} = \dfrac{\square}{\square}$    **4** $\dfrac{24}{40} = \dfrac{\square}{\square}$

**What You Will Learn**

*Lesson 4-2*

The slope describes the "steepness" of the line as the ratio of rise over run.

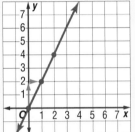

$$\text{slope} = \dfrac{\text{rise}}{\text{run}} = \dfrac{2}{1}$$

Rise describes the vertical change. Run describes the horizontal change. The slope of a line is written in simplest form.

---

You know how to graph an ordered pair.

**Example:** Point $A$ is at $(1, 3)$.

**TRY IT!**

**5** Graph and label point $B(3, 4)$ on the coordinate grid.

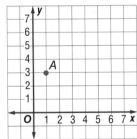

*Lesson 4-4*

You can graph linear equations using a table.

Graph $y = 2x$ using a table.

| x | y = 2x | y |
|---|---|---|
| 0 | 2(0) | 0 |
| 1 | 2(1) | 2 |
| 2 | 2(2) | 4 |

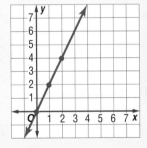

# Relations and Functions

## KEY Concept

An **ordered pair** is a set of numbers that identifies a point on a coordinate grid.

$x$-coordinate ⎯⎯⎯⎯⎯⎯ $y$-coordinate

A **relation** is a set of ordered pairs.

$\{(1, 6), (7, -3), (-2, 3), (0, 9), (-5, 0)\}$

The **domain** is the set of $x$-coordinates. Typically, domain is listed in order from least to greatest.      $\{-5, -2, 0, 1, 7\}$

The **range** is the set of $y$-coordinates. Typically, range is listed in order from least to greatest.      $\{-3, 0, 3, 6, 9\}$

Some relations are functions. A **function** is a relation in which each domain value is paired with exactly one range value.

Relations are shown as a list, as a table, as a mapping, or as a graph. A relation that is a function is shown in the table below.

| x | −5 | −2 | 0 | 1 | 7 |
|---|---|---|---|---|---|
| y | 0 | 3 | 9 | 6 | −3 |

The relation is a function because each domain value ($x$) is paired with exactly one range value ($y$).

VOCABULARY

**domain**
the set of first numbers of the ordered pairs in a relation

**function**
a relation in which each element of the domain is paired with exactly one element of the range

**ordered pair**
a set of numbers or coordinates used to locate any point on a coordinate plane, written in the form $(x, y)$

**range**
the set of second numbers of the ordered pairs in a relation

**relation**
a set of ordered pairs

Not every relation is a function.

## Example 1

**State the domain and range of the relation.**

1. The domain is the set of $x$-values.
   domain = {0, 1, 4}

2. The range is the set of $y$-values.
   range = {3, 5, 11}

| Domain | Function | Range |
|---|---|---|
| x | y = 2x + 3 | y |
| 0 | y = 2(0) + 3 | 3 |
| 1 | y = 2(1) + 3 | 5 |
| 4 | y = 2(4) + 3 | 11 |

## YOUR TURN!

**State the domain and range of the relation.**

1. The domain is the set of _____.

   domain = {_____}

2. The range is the set of _____.

   range = {_____}

| Domain | Function | Range |
|:------:|:--------:|:-----:|
| **x** | **y = −x − 1** | **y** |
| −3 | y = −(−3) − 1 | 2 |
| 0 | y = −(0) − 1 | −1 |
| 1 | y = −(1) − 1 | −2 |

---

## Example 2

**State the domain and range of the relation.**

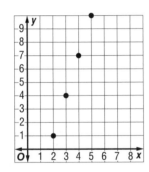

1. The ordered pairs are $(2, 1)$, $(3, 4)$, $(4, 7)$, $(5, 10)$.

2. The domain is the set of $x$-values.
   **domain = {2, 3, 4, 5}**

3. The range is the set of $y$-values.
   **range = {1, 4, 7, 10}**

## YOUR TURN!

**State the domain and range of the relation.**

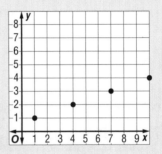

1. The ordered pairs are (___, ___), (___, ___), (___, ___), (___, ___).

2. The domain is the set of _____.

   domain = {_____}

3. The range is the set of _____.

   range = {_____}

GO ON

## Example 3

**State the domain and range of the relation. Tell whether the relation is a function.**

$$\{(-3, 10), (-1, 0), (-1, 3), (2, -5)\}$$

1. The domain is the set of $x$-values.
   **domain** = {−3, −1, 2}

2. The range is the set of $y$-values.
   **range** = {−5, 0, 3, 10}

3. Name any domain values that have more than one range value. −1 is paired with both 0 and 3.
   So, the relation is not a function.

**YOUR TURN!**

**State the domain and range of the relation. Tell whether the relation is a function.**

$$\{(1, 4), (-5, -2), (8, 2), (5, 12)\}$$

1. The domain is the set of _____.

   domain = {_____}

2. The range is the set of _____.

   range = {_____}

3. Name any domain values that have more than one range value. _____

   So, the relation _____ a function.

## ▶ Guided Practice

**State the domain and range of each relation.**

**1**

| x | y = x − 1 | y |
|---|---|---|
| −5 | y = (−5) − 1 | −6 |
| −2 | y = (−2) − 1 | −3 |
| 1 | y = (1) − 1 | 0 |

The domain is the set of _____.

domain = {_____}

The range is the set of _____.

range = {_____}

**2**

| x | y = 7x + 3 | y |
|---|---|---|
| 2 | y = 7(2) + 3 | 17 |
| 4 | y = 7(4) + 3 | 31 |
| 5 | y = 7(5) + 3 | 38 |

domain = {_____}

range = {_____}

**3**

| x | y = 2x + 3 | y |
|---|---|---|
| −2 | y = 2(−2) + 3 | −1 |
| 1 | y = 2(1) + 3 | 5 |
| 4 | y = 2(4) + 3 | 11 |

domain = {_____}

range = {_____}

**4**

| x | y = −x + 1 | y |
|---|---|---|
| −3 | y = −(−3) + 1 | 4 |
| 0 | y = −(0) + 1 | 1 |
| 5 | y = −(5) + 1 | −4 |

domain = {_____}

range = {_____}

**5**  State the domain and range of the relation.

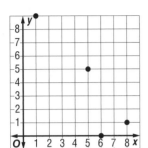

**Step 1**  The ordered pairs are (_____, _____),(_____, _____),

(_____, _____),(_____, _____).

**Step 2**  The domain is the set of _____.

The domain = {_____}

**Step 3**  The range is the set of _____.

The range = {_____}

**State the domain and range of the relation.**

**6**

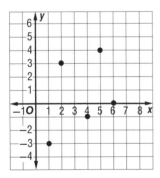

domain = _____

range = _____

**7**

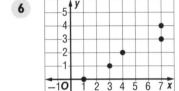

domain = _____

range = _____

**State the domain and range of each relation. Tell whether each relation is a function.**

**8**  {(−3, −14), (4, 6), (5, 18), (12, 6)}

domain = _____

range = _____

Do any domain values have more

than one range value? _____

The relation _____ a function.

**9**  {(3, 5), (5, 1), (9, 0), (1, 2)}

domain = _____

range = _____

Do any domain values have more

than one range value? _____

The relation _____ a function.

**Solve.**

**10  LANDSCAPING** Farah is buying bags of mulch for landscaping. The price for 15 bags is $75. The price for 25 bags is $125. The price for 40 bags is $200, and the price for 50 bags is $250. Name the domain of quantities and range of prices from which Farah has to choose.

| Quantity of Bags, $x$ | Price, $y$ |
|---|---|
|  |  |
|  |  |
|  |  |
|  |  |

The $x$-values represent _____.

The $y$-values represent _____.

Check off each step.

_____ Understand: I underlined key words.

_____ Plan: To solve the problem, I will _____.

_____ Solve: The answer is _____

_____.

_____ Check: I checked my answer by _____

_____.

 **Skills, Concepts, and Problem Solving**

**State the domain and range of each relation.**

**11**

| $x$ | $y = 2x + 2$ | $y$ |
|---|---|---|
| −1 | $y = 2(-1) + 2$ | 0 |
| 0 | $y = 2(0) + 2$ | 2 |
| 2 | $y = 2(2) + 2$ | 6 |
| 7 | $y = 2(7) + 2$ | 16 |

domain = _____

range = _____

**12**

| $x$ | $y = 4x - 2$ | $y$ |
|---|---|---|
| −2 | $y = 4(-2) - 2$ | −10 |
| 3 | $y = 4(3) - 2$ | 10 |
| 7 | $y = 4(7) - 2$ | 26 |

domain = _____

range = _____

**State the domain and range of the relation. Tell whether the relation is a function.**

**13**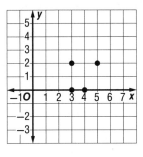

domain = _____

range = _____

The relation _____ a function.

**14**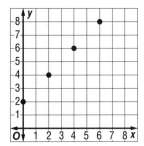

domain = _____

range = _____

The relation _____ a function.

**15** {(1, 3), (3, 10), (2, 6), (2, 4)}

domain = _____

range = _____

The relation _____ a function.

**16** {(2, 6), (1, 12), (0, 0), (−1, 9)}

domain = _____

range = _____

The relation _____ a function.

**17**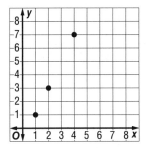

ordered pairs {(_____, _____), (_____, _____), (_____, _____)}

domain = {_____}

range = {_____}

The relation _____ a function.

**18**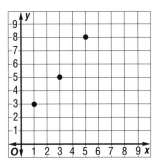

ordered pairs {(_____, _____), (_____, _____), (_____, _____)}

domain = {_____}

range = {_____}

The relation _____ a function.

**Solve.**

**19** **CEMENT** The directions on the side of a bag of cement mix shows a table with the amount of water that should be used with each amount of cement mix.

| Cement mix (part of bag) | $\frac{1}{4}$ | $\frac{1}{2}$ | 1 |
|---|---|---|---|
| Liters of water | 7 | 14 | 28 |

Let the domain of this relation be the amount of cement mix and the range be the liters of water. What are the domain and the range?

_____

_____

**20** **BASEBALL** A baseball pitcher keeps track of how many pitches he throws and how many runs he gives up in each game. The table below shows the data. If the pitches thrown are the domain, and the range is the number of runs he allowed, what are the domain and range of the relation?

| Pitches Thrown | 78 | 95 | 89 | 101 |
|---|---|---|---|---|
| Runs Allowed | 3 | 1 | 0 | 5 |

_____

_____

**Vocabulary Check** **Write the vocabulary word that completes each sentence.**

**21** (2, 7) is a(n) _____.

**22** The set of *x*-values in a relation is the _____.

**23** A(n) _____ is a relation in which each element of the domain is paired with exactly one element of the range.

**24** **Reflect** Can a domain have the same numbers as a range? If so, give an example.

_____

_____

STOP

## KEY Concept

The **slope** of a **linear function** describes the "steepness" of the line. It is the ratio of the rise over the run.

$$\text{slope} = \frac{\text{rise}}{\text{run}} = \frac{1}{2}$$

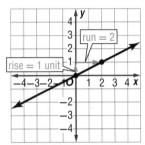

Another way to find slope is to use ordered pairs. The rise is the change in the $y$-values, and the run is the change in the $x$-values.

The ordered pairs $(0, 0)$ and $(2, 1)$ describe points on the line graphed above. Use these points in the formula.

$$(x_1, y_1) \quad (x_2, y_2)$$
$$(0, 0) \quad (2, 1)$$

$$\text{slope} = \frac{y_2 - y_1}{x_2 - x_1} = \frac{1 - 0}{2 - 0} = \frac{1}{2}$$

Lines that move upward and to the right have a positive slope. Lines that move downward and to the left have a negative slope.

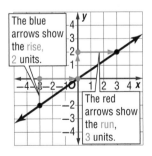

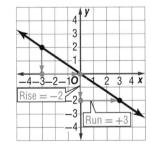

$$\frac{\text{rise}}{\text{run}} = \frac{+2}{+3} = \text{positive slope} \qquad \frac{\text{rise}}{\text{run}} = \frac{-2}{+3} = \text{negative slope}$$

A negative slope can have a negative rise and a positive run, or a positive rise and a negative run.

### VOCABULARY

**linear function**
a function with ordered pairs that satisfy a linear equation

**slope**
the ratio of the change in the $y$-coordinates (rise) to the corresponding change in the $x$-coordinates (run) as you move from one point to another along a line

## Example 1

**Find the slope of the line.**

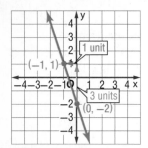

1. Identify two points on the line.
   $(-1, 1)$ and $(0, -2)$ are on the line.

2. Count the rise of the line from
   $-2$ to $1$. The rise is $+3$.

3. Count the run of the line from $0$ to $-1$.
   The run is $-1$.

4. Write the slope as a ratio.

$$\text{slope} = \frac{\text{rise}}{\text{run}} = \frac{3}{-1} = -3$$

YOUR TURN!

**Find the slope of the line.**

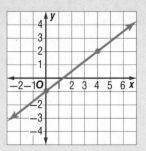

1. Identify two points on the line.

   (_____, _____) and (_____, _____)
   are on the line.

2. Count the rise of the line from

   _____ to _____. The rise is _____.

3. Count the run of the line from _____

   to _____. The run is _____.

4. Write the slope as a ratio.

$$\text{slope} = \frac{\text{rise}}{\text{run}} = \frac{\boxed{\phantom{x}}}{\boxed{\phantom{x}}}$$

## Example 2

**Find the slope of the line that contains the points $(-3, 6)$ and $(2, 4)$.**

1. Label the points.

   $(x_1, y_1)$      $(x_2, y_2)$

   $(-3, 6)$      $(2, 4)$

2. Substitute values into the slope formula.

$$\text{slope} = \frac{y_2 - y_1}{x_2 - x_1}$$

$$\frac{4 - 6}{2 - (-3)} = \frac{-2}{5} = -\frac{2}{5}$$

YOUR TURN!

**Find the slope of the line that contains the points $(4, -1)$ and $(8, 11)$.**

1. Label the points.

   $(x_1, y_1)$      $(x_2, y_2)$

   (_____, _____)    (_____, _____)

2. Substitute values into the slope formula.

$$\text{slope} = \frac{y_2 - y_1}{x_2 - x_1}$$

$$\frac{\boxed{\phantom{x}} - \boxed{\phantom{x}}}{\boxed{\phantom{x}} - \boxed{\phantom{x}}} = \frac{\boxed{\phantom{x}}}{\boxed{\phantom{x}}} = \underline{\phantom{xxx}}$$

## ▶ Guided Practice

**Find the slope of each line.**

**1**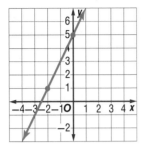

Identify two points on the line.

(−2, _____) and (_____, 5)

The rise is _____.

The run is _____.

$$\text{slope} = \frac{\text{rise}}{\text{run}} = \frac{\boxed{\phantom{xx}}}{\boxed{\phantom{xx}}} = \text{_____}$$

**2**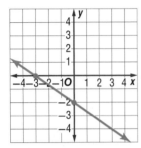

Identify two points on the line.

(_____, _____) and (_____, _____)

The rise is _____.

The run is _____.

$$\text{slope} = \frac{\text{rise}}{\text{run}} = \frac{\boxed{\phantom{xx}}}{\boxed{\phantom{xx}}} = \text{_____}$$

---

## Step by Step Practice

**3** Find the slope of the line that contains the points (−4, 0) and (3, −2).

**Step 1** Label the points.

$(x_1, y_1) \qquad (x_2, y_2)$

(_____, _____)   (_____, _____)

**Step 2** Substitute values into the slope formula.

$$\text{slope} = \frac{y_2 - y_1}{x_2 - x_1} = \text{_____} = \text{_____}$$

GO ON

**Find each slope of the line that contains each pair of points.**

**4** $(3, 1)$ and $(7, -7)$

$(x_1, y_1)$ $\qquad\qquad$ $(x_2, y_2)$

_____ $\qquad\quad$ _____

Substitute values into the slope formula.

$$\text{slope} = \frac{y_2 - y_1}{x_2 - x_1} = \frac{\boxed{\phantom{x}} - \boxed{\phantom{x}}}{\boxed{\phantom{x}} - \boxed{\phantom{x}}} = \frac{\boxed{\phantom{x}}}{\boxed{\phantom{x}}} = \underline{\phantom{xxx}}$$

**5** $(0, -1)$ and $(4, -5)$

Substitute values into the slope formula.

$$\text{slope} = \frac{y_2 - y_1}{x_2 - x_1} = \underline{\phantom{xxxxxxxxx}} = \underline{\phantom{xxx}} = \underline{\phantom{xx}}$$

## Step by Step Problem-Solving Practice

**Solve.**

**6** **TICKET SALES**   Kareem plots the points $(1, 5)$ and $(3, 15)$ on a graph to find the line that represents the number of dollars he earns by selling tickets to a raffle. What is the slope of the line? What does the slope mean to Kareem?

$(x_1, y_1)$ $\qquad\qquad$ $(x_2, y_2)$

_____ $\qquad\quad$ _____

$$\text{slope} = \underline{\phantom{xxxxxx}} = \underline{\phantom{xxxxxx}} = \underline{\phantom{xx}} = \underline{\phantom{xx}}$$

Check off each step.

_____ **Understand: I underlined key words.**

_____ **Plan: To solve the problem, I will** _____.

_____ **Solve: The answer is** _____.

_____ **Check: I checked my answer by** _____

_____.

**Skills, Concepts, and Problem Solving**

**Find the slope of each line.**

**7**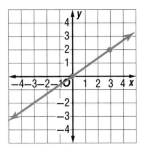

_____

**8**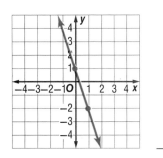

_____

**Find the slope of the line that contains each pair of points.**

**9** (2, 7) and (5, 3)  _____

**10** (2, 1) and (7, 4)  _____

**11** (0, 5) and (−5, 0) _____

**12** (−3, 3) and (5, −4) _____

**Solve.**

**13** **ROADS**   A city planner wants to know the rate of change in elevation of Maple Road. A sketch names the points (100, 5) and (225, 15) as being on that road. What is the slope of the road?

_____

**14** **LANDSCAPING**   The landscaper plans to change the slope of the hill in Helen's backyard. A sketch labels the point at the top of the hill (2, 15). The point at the bottom of the hill is labeled (3, 3). What is the slope of the hill as shown on the sketch?

_____

**Vocabulary Check**   **Write the vocabulary word that completes each sentence.**

**15** A(n) _____ is a function with ordered pairs that satisfy a linear equation.

**16** The ratio of the change in $y$-values to the corresponding changes

in $x$-values is the _____ of a linear function.

**17** **Reflect**   Can you use any two points on a line to determine the slope of the line? Does it matter which point you use as $(x_1, y_1)$? Explain.

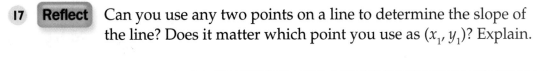

**Identify the domain and range of each relation. Tell whether the relation is a function.**

**1**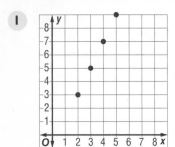

domain = _____

range = _____

The relation _____ a function.

**2**

| $x$ | $y = -x - 1$ | $y$ |
|-----|--------------|-----|
| −4 | $y = -(-4) - 1$ | 3 |
| 0 | $y = -(0) - 1$ | −1 |
| 1 | $y = -(1) - 1$ | −2 |
| 5 | $y = -(5) - 1$ | −6 |

domain = _____

range = _____

The relation _____ a function.

**Find the slope of each line.**

**3**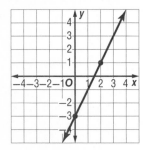

_____

**4**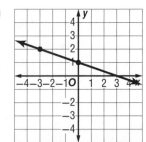

_____

**Find the slope of the line that contains each pair of points.**

**5** (−3, −5) and (3, 3) _____

**6** (0, −2) and (1, 1) _____

**Solve.**

**7** **PETS** At the pet store, the sign above the fish tank gave the prices as 3 fish for $6 and 5 fish for $10. At home Alicia made a graph for the prices of fish to decide how many fish she could afford to buy. Find the slope of the line to determine the price for each fish.

_____

# Slope-Intercept Form

## KEY Concept

**Linear equations** can be written in different forms. Slope-intercept form is the most commonly used form

**Slope-Intercept Form**

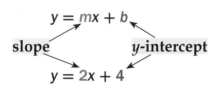

$$y = mx + b$$

slope          y-intercept

$$y = 2x + 4$$

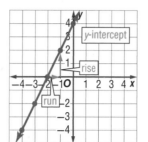

You can use a graph to find the equation of a line. First, locate the y-intercept and then determine the slope.

The line crosses the y-axis at (0, 4).

$$\text{slope} = \frac{\text{rise}}{\text{run}} = \frac{2}{1} = 2$$

### VOCABULARY

**linear equation**
> an equation with a graph that is a straight line

**slope (*m*)**
> the ratio of the change in the y-value to the corresponding change in the x-value in a linear function

**y-intercept (*b*)**
> the y-coodinate of the point at which a graph crosses the y-axis

Remember that the slope is the ratio of the rise over the run. The slope can also be found using the slope formula.

## Example 1

**Write $2y = 3x + 4$ in slope-intercept form. Name the slope and y-intercept.**

1. Solve the equation for y.

$$\frac{2y}{2} = \frac{3x}{2} + \frac{4}{2} \qquad \text{Divide each side by 2.}$$

$$y = \frac{3}{2}x + 2$$

2. The slope is the coefficient of x.

The slope is $\frac{3}{2}$.

3. The y-intercept is the constant. The y-intercept is 2, which means the line passes through (0, 2).

## YOUR TURN!

**Write $y + 6x = 3$ in slope-intercept form. Name the slope and y-intercept.**

1. Solve the equation for y.

$$y + 6x = 3 \qquad \underline{\hspace{2cm}}.$$

$$\underline{\hspace{4cm}}$$

$$y = \underline{\hspace{3cm}}$$

2. The slope is the $\underline{\hspace{4cm}}$.

The slope is $\underline{\hspace{1cm}}$.

3. The y-intercept is the $\underline{\hspace{2.5cm}}$. The y-intercept is $\underline{\hspace{1cm}}$, which means the line passes through $\underline{\hspace{2cm}}$.

**GO ON**

## Example 2

Identify the equation of the line. Write the equation in slope-intercept form.

1. Find the $y$-intercept.

   $(0, -5)$  $b = -5$

2. Find the slope.

   $\text{slope} = \dfrac{\text{rise}}{\text{run}} = \dfrac{5}{5} = 1$

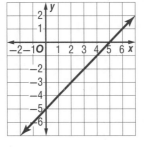

3. Write the equation.

   $y = mx + b$  Slope-intercept form

   $y = 1x + (-5)$  Substitute.

   $y = x - 5$  Simplify.

## YOUR TURN!

Identify the equation of the line. Write the equation in slope-intercept form.

1. Find the $y$-intercept.

   $(0, \underline{\hphantom{xx}})$  $b = \underline{\hphantom{xx}}$

2. Find the slope.

   $\text{slope} = \dfrac{\text{rise}}{\text{run}} = \dfrac{\square}{\square}$

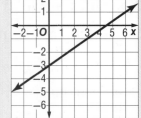

3. Write the equation.

   $y = \underline{\hphantom{xx}} x + \underline{\hphantom{xx}}$  Slope-intercept form

   $y = \dfrac{\square}{\square} x + (\underline{\hphantom{xx}})$  _____

   $y = \underline{\hphantom{xx}}$  _____

## Example 3

Write the equation of the line that passes through the point (3, 2) and has a slope of −2.

1. Use the slope-intercept form and solve for $b$. Substitute for $x$, $y$, and $m$.

   $x = 3, y = 2, m = -2$

   $y = mx + b$

   $2 = -2(3) + b$  Substitute.

   $2 = -6 + b$  Simplify.

   $8 = b$

2. Substitute the values for $m$ and $b$ into the slope-intercept form to write the equation of the line.

   $m = -2, b = 8$

   $y = mx + b$

   $y = -2x + 8$  Substitute.

## YOUR TURN!

Write the equation of the line that passes through the point (−1, −3) and has a slope of 4.

1. Use the slope-intercept form and solve for $b$. Substitute for $x$, $y$, and $m$.

   $x = \underline{\hphantom{xx}}, y = \underline{\hphantom{xx}}, m = \underline{\hphantom{xx}}$

   $y = mx + b$

   $\underline{\hphantom{xx}} = \underline{\hphantom{xx}}(\underline{\hphantom{xx}}) + b$  _____

   $\underline{\hphantom{xx}} = \underline{\hphantom{xx}} + b$  _____

   $\underline{\hphantom{xx}} = b$

2. Substitute the values for $m$ and $b$ into the slope-intercept form to write the equation of the line.

   $m = \underline{\hphantom{xx}}, b = \underline{\hphantom{xx}}$

   $y = mx + b$

   $y = \underline{\hphantom{xx}} x + \underline{\hphantom{xx}}$  _____

# Guided Practice

**Name the slope and *y*-intercept for each equation.**

**1** $y = 4x + 3$

$m =$ _____

$y$-intercept $(0,$ _____$)$

**2** $x + 3y = -12$

$m =$ _____

$y$-intercept $(0,$ _____$)$

**Write each equation in slope-intercept form. Name the slope and the *y*-intercept.**

**3** $2x - 5y = -5$

_____

_____

_____

_____

$m =$ _____

$b =$ _____

The line passes through

the point $(0,$ _____$)$

**4** $4x = -y + 9$

_____

_____

_____

_____

$m =$ _____

$b =$ _____

The line passes through

the point $(0,$ _____$)$

**Identify the equation of each line. Write the equation in slope-intercept form.**

**5**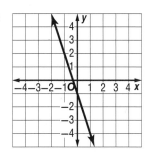

$m =$ _____

$b =$ _____

$y =$ _____ $x + ($ _____$)$

**6**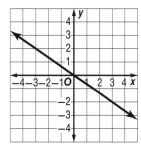

$m =$ _____

$b =$ _____

$y =$ _____

GO ON

Copyright © Glencoe/McGraw-Hill, a division of The McGraw-Hill Companies, Inc.

**Lesson 4-3** Slope-Intercept Form  **123**

**7** Write the equation of the line that passes through the point $(2, -1)$ and has a slope of $\frac{1}{3}$.

**Step 1** Use the slope-intercept form and solve for $b$. Substitute for $x$, $y$, and $m$.

$$y = mx + b \qquad x = \underline{\hspace{1cm}}, y = \underline{\hspace{1cm}}, m = \underline{\hspace{1cm}}$$

$$\underline{\hspace{1cm}} = \underline{\hspace{1cm}}(\underline{\hspace{1cm}}) + b$$

$$\underline{\hspace{1cm}} = \underline{\hspace{1cm}} + b$$

$$\underline{\hspace{2cm}} = b$$

$$\underline{\hspace{2cm}} = b$$

$$\underline{\hspace{2cm}} = b$$

**Step 2** Substitute the values for $m$ and $b$ into the slope-intercept form to write the equation of the line.

$$y = mx + b \qquad m = \underline{\hspace{1cm}}, b = \underline{\hspace{1cm}}$$

$$y = \underline{\hspace{1cm}}x + (\underline{\hspace{1cm}})$$

**Write the equation for the line that passes through the given point and has the given slope.**

**8** $(6, 0), m = -1$

$$x = \underline{\hspace{1cm}}, y = \underline{\hspace{1cm}}, m = \underline{\hspace{1cm}}$$

$$y = mx + b$$

$$\underline{\hspace{1cm}} = \underline{\hspace{1cm}}(\underline{\hspace{1cm}}) + b$$

$$\underline{\hspace{1cm}} = \underline{\hspace{1cm}} + b$$

$$\underline{\hspace{1cm}} = b$$

$$y = \underline{\hspace{1cm}}x + (\underline{\hspace{1cm}})$$

**9** $(5, 4), m = \frac{1}{5}$

$$x = \underline{\hspace{1cm}}, y = \underline{\hspace{1cm}}, m = \underline{\hspace{1cm}}$$

$$y = mx + b$$

$$\underline{\hspace{1cm}} = \underline{\hspace{1cm}}(\underline{\hspace{1cm}}) + b$$

$$\underline{\hspace{1cm}} = \underline{\hspace{1cm}} + b$$

$$\underline{\hspace{1cm}} = b$$

$$y = \underline{\hspace{1cm}}x + (\underline{\hspace{1cm}})$$

**Solve.**

**10** **TERM PAPER** Javier is working on a term paper. He has been working $x$ hours. The equation $2y - x = 10$ represents the number of pages completed. What is the slope of this equation? What is the $y$-intercept?

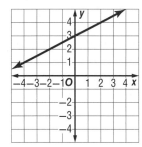

Solve the equation for $y$.

$$2y - x = 10$$

_____ Add $x$ to each side of the equation.

_____ Divide each side by 2 to isolate $y$.

Check off each step.

_____ Understand: I underlined key words.

_____ Plan: To solve the problem, I will _____.

_____ Solve: The answer is _____.

_____ Check: I checked my answer by _____.

## Skills, Concepts, and Problem Solving

**Identify the equation of each line. Write the equation in slope-intercept form.**

**11**

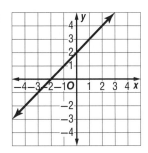

$m = $ _____

$b = $ _____

$y = $ _____

**12**

$m = $ _____

$b = $ _____

$y = $ _____

GO ON

**Write the equation of each line in slope-intercept form.**

**13** $-8x = 14y$

**14** $9y + 6x = 45$

**Write the equation for the line that passes through each given point and has each given slope.**

**15** $(8, 0),\ m = \dfrac{1}{4}$

**16** $(12, 17),\ m = \dfrac{1}{3}$

**17** $(-2, 2),\ m = \dfrac{1}{2}$

**18** $(-3, -3),\ m = -1$

**Solve.**

**19** **TRAVELING**  Alexa went on a car trip with her family. She records the number of hours and the number of miles they have traveled on a coordinate grid. Two of the points she recorded are $(2, 110)$ and $(7, 385)$. Assume they are traveling at a constant rate. Write an equation to represent her family trip.

_____

**20** **GRAPHING**  Parallel lines have equal slopes. Write the equation of a line parallel to the line graphed at the right. The $y$-intercept of the parallel line is $(0, -11)$.

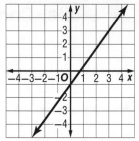

_____

**Vocabulary Check**  **Write the vocabulary word that completes each sentence.**

**21** The $y$-coordinate of the point at which a graph crosses the $y$-axis is

called the _____.

**22** The _____ is the ratio of change in the $y$-value to the corresponding change in the $x$-value in a linear function.

**23** **Reflect**  In Exercise 19, what does the ordered pair $(2, 110)$ represent?

_____

_____

# Graph Linear Equations

## KEY Concept

To graph a line given in **slope-intercept form**, first plot the **y-intercept** (*b*) and then use the **slope** (*m*) to plot another point. Connect the points to draw the line.

$$y = mx + b$$

$$y = \frac{2}{3}x - 1$$

The equation $y = \frac{2}{3}x + (-1)$ is rewritten as $y = \frac{2}{3}x - 1$.

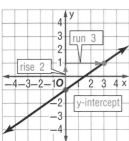

### VOCABULARY

**slope** *(m)*
the ratio of the change in the *y*-value to the corresponding change in the *x*-value in a linear function

**slope-intercept form**
an equation of the form $y = mx + b$, where *m* is the slope and *b* is the *y*-intercept

**y-intercept** *(b)*
the *y*-coordinate of the point at which a graph crosses the *y*-axis

---

## Example 1

**Graph $y = \frac{1}{2}x - 1$ using a table.**

1. Complete the table. Substitute values for *x*. Solve for *y*.

| x | $y = \frac{1}{2}x - 1$ | y |
|---|---|---|
| −2 | $y = \frac{1}{2}(-2) - 1$ | −2 |
| 0 | $y = \frac{1}{2}(0) - 1$ | −1 |
| 2 | $y = \frac{1}{2}(2) - 1$ | 0 |

2. Plot the points on a graph.

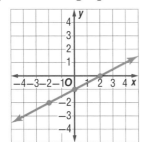

3. Connect the points with a line.

## YOUR TURN!

**Graph $y = -2x + 1$ using a table.**

1. Complete the table. Substitute values for *x*. Solve for *y*.

| x | $y = -2x + 1$ | y |
|---|---|---|
| −1 | $y = -2(\underline{\quad}) + 1$ | |
| 0 | $y = -2(\underline{\quad}) + 1$ | |
| 1 | $y = -2(\underline{\quad}) + 1$ | |

2. Plot the points on a graph.

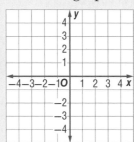

3. Connect the points with a line.  **GO ON**

## Example 2

**Graph the equation $y = -\frac{1}{3}x + 1$.**

1. slope $= \dfrac{\text{rise}}{\text{run}} = \dfrac{-1}{3}$

2. $y$-intercept $= 1$

   Plot the point for the $y$-intercept, $(0, 1)$.

3. From that point, rise $-1$ and run $3$ to plot another point.

4. Draw a line through the points.

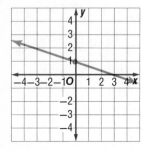

## YOUR TURN!

**Graph the equation $y = x - 2$.**

1. slope $= \dfrac{\text{rise}}{\text{run}} = \dfrac{\square}{\square}$    When the slope is a whole number, think of it as a fraction over 1.

2. $y$-intercept $=$ _____

   Plot the point for the $y$-intercept,

   (_____ , _____).

3. From that point, rise _____ and run _____ to plot another point.

4. Draw a line through the points.

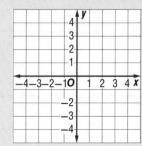

## ▶ Guided Practice

**Graph each equation using a table.**

1  $y = -2x - 1$

| x | y = −2x −1 | y |
|----|----|----|
| −2 | −2 (___) − 1 | |
| 0 | −2 (___) − 1 | |
| 2 | −2 (___) − 1 | |

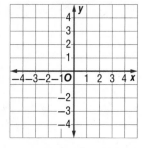

2  $y = -\frac{2}{3}x + 2$

| x | $y = -\frac{2}{3}x + 2$ | y |
|----|----|----|
| −3 | $-\frac{2}{3}$ (___) + 2 | |
| 0 | $-\frac{2}{3}$ (___) + 2 | |
| 3 | $-\frac{2}{3}$ (___) + 2 | |

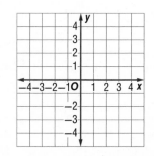

**Graph each equation.**

**3** $y = 4x - 1$

slope $= \dfrac{\text{rise}}{\text{run}} = \dfrac{\boxed{\phantom{x}}}{\boxed{\phantom{x}}}$

The $y$-intercept is (0, _____).

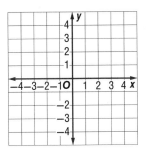

**4** $y = -\dfrac{3}{4}x + 3$

slope $= \dfrac{\text{rise}}{\text{run}} = \dfrac{\boxed{\phantom{x}}}{\boxed{\phantom{x}}}$

The $y$-intercept is (0, _____).

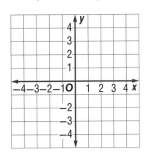

**Step by Step Practice**

**5** Graph the equation $y = \dfrac{2}{3}x$.

**Step 1** Complete the table. Substitute the values for $x$. Solve for $y$.

| $x$ | $y = \dfrac{2}{3}x$ | $y$ |
|---|---|---|
| −3 | $y = \dfrac{2}{3}(\underline{\phantom{xx}})$ | |
| 0 | $y = \dfrac{2}{3}(\underline{\phantom{xx}})$ | |
| 3 | $y = \dfrac{2}{3}(\underline{\phantom{xx}})$ | |

**Step 2** Plot the points on a graph.

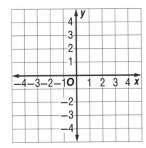

**Step 3** Connect the points with a line.

GO ON

**Graph each equation.**

**6** $y = -\frac{4}{3}x - 1$

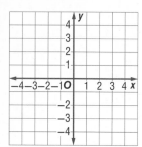

The slope is _____

The $y$-intercept is $(0, \underline{\quad})$.

**7** $y = \frac{3}{4}x + 1$

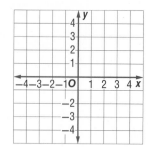

The slope is _____

The $y$-intercept is $(\underline{\quad}, \underline{\quad})$.

## Step by Step Problem-Solving Practice

**Solve.**

**8** **FINANCIAL LITERACY** Dan earns $5 every week as an allowance and then earns $2 for each hour he spends on chores. The equation that models how much money he earns each week is $y = 2x + 5$, where $y$ is the total money for the week and $x$ is the number of hours he completes chores. How much will Dan earn if he works for 5 hours on chores in one week?

$y = 2x + 5$     The slope is _____.

The $y$-intercept is _____.

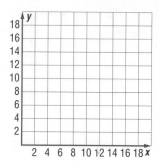

Check off each step.

_____ Understand: I underlined key words.

_____ Plan: To solve the problem, I will _____.

_____ Solve: The answer is _____.

_____ Check: I checked my answer by _____

_____.

# ▶ Skills, Concepts, and Problem Solving

**Graph each equation.**

**9** $y = \dfrac{1}{3}x$

$m = $ _____

$b = $ _____

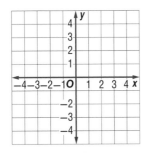

**10** $y = \dfrac{1}{4}x - 4$

$m = $ _____

$b = $ _____

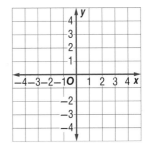

**11** $y = -2x + 2$

$m = $ _____

$b = $ _____

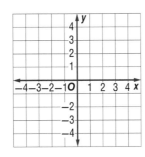

**12** $y = 3x + 5$

$m = $ _____

$b = $ _____

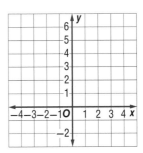

**Solve.**

**13** **SALES**  Tabitha is having a lemonade sale in her neighborhood. She spent $20 to buy several cans of lemonade and is selling the cans for $1 each. Her profit is modeled by the equation $y = x - 20$, where $y$ is her profit and $x$ represents the number of cans she sells. Graph this equation.

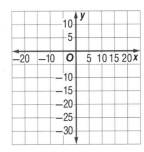

**Vocabulary Check**  **Write the vocabulary word that completes each sentence.**

**14** In the slope-intercept form of a line, $b$ represents the _____.

**15** The slope-intercept form of a line is _____.

**16** **Reflect**  You are given one point on a line. What other information do you need to be able to graph the line?

_____

_____

_____

_____

STOP

**Write each equation in slope-intercept form. Name the slope and the y-intercept.**

**1** $3y - 9 = 2x$

$m =$ _____  $b =$ _____

**2** $2y + 2 = 2x$

$m =$ _____  $b =$ _____

**Write the equation for the line that passes through each given point and has each given slope.**

**3** $(0, 0), m = \dfrac{1}{2}$

**4** $(4, 1), m = 1$

**Graph each equation.**

**5** $y = \dfrac{1}{4}x$

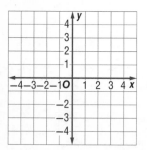

**6** $y = -3x + 1$

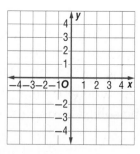

**7** $y = 2x + 3$

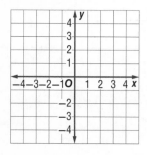

**8** $y = x - 5$

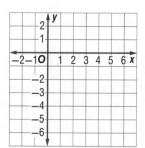

**Solve. Write the answer in simplest terms.**

**9** **SAVINGS**  Jenna opened a savings account with a $75 deposit. Every week she puts $50 in her account. Write an equation for the amount of money in her account after $x$ weeks.

_____

# Solve Systems of Linear Equations Using Graphs

## KEY Concept

A **system of equations** is a set of two or more equations with the same variables. The **solution of a system of equations** occurs where the graphs of the lines intersect.

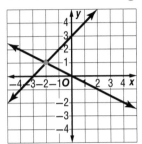
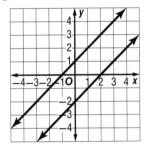
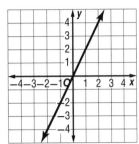

**One Solution**
Two different lines intersect in one point

**No Solutions**
Two parallel lines do not intersect

**Infinite Solutions**
Two lines that coinside intersect at every point on on the lines

VOCABULARY

**linear function**
a function with ordered pairs that satisify a linear equation

**solution of a system of equations**
an ordered pair that satisfies both equations

**system of equations**
a set of equations with the same variables

To find the solution of a system of linear equations, graph both equations on the same coordinate grid.

## Example 1

**Solve the system of equations by graphing.**
$$y = 3x - 2$$
$$y = -x - 6$$

1. Graph $y = 3x - 2$.

$m = \dfrac{3}{1}$    $b = -2$    $y$-intercept: $(0, -2)$

2. Graph $y = -x - 6$ on the same grid.

$m = -\dfrac{1}{1}$    $b = -6$    $y$-intercept: $(0, -6)$

3. The lines intersect at $(-1, -5)$.

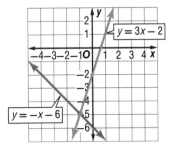

## YOUR TURN!

**Solve the system of equations by graphing.**
$$y = 4x + 2$$
$$y = 3x + 3$$

1. Graph $y = 4x + 2$.

$m = $____    $b = $____    $y$-intercept: $(0, $____$)$

2. Graph $y = 3x + 3$ on the same grid.

$m = $____    $b = $____    $y$-intercept: $(0, $____$)$

3. The lines intersect at _____.

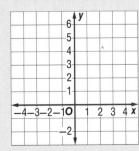

GO ON

## Example 2

**Solve the system of equations by graphing.**
$$-2x + y = -1$$
$$2y - 4x = 6$$

1. Write each equation in slope-intercept form.

   $-2x + y = -1$      $2y - 4x = 6$

         $y = 2x - 1$       $2y = 4x + 6$

                            $y = 2x + 3$

2. Graph $-2x + y = -1$.

   $m = \dfrac{2}{1}$     $b = -1$

   $y$-intercept: $(0, -1)$

3. Graph $2y - 4x = 6$.

   $m = \dfrac{2}{1}$     $b = 3$

   $y$-intercept: $(0, 3)$

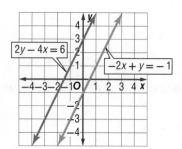

4. The lines are parallel. There is no solution to this system.

## YOUR TURN!

**Solve the system of equations by graphing.**
$$-x + 3y = 2$$
$$-3x + 9y = 6$$

1. Write both equations in slope-intercept form.

   $-x + 3y = 2$      $-3x + 9y = 6$

       $3y =$ _____ $+ 2$      $9y =$ _____ $+ 6$

            _____           _____

2. Graph $x - 3y = 2$.

   $m =$ _____     $b =$ _____

   $y$-intercept: $(0,$ _____$)$

3. Graph $-3x + 9y = 6$.

   $m =$ _____     $b =$ _____

   $y$-intercept: $(0,$ _____$)$

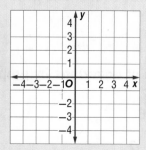

4. The lines are _____

   _____

## Guided Practice

**Solve the system of equations by graphing.**

1.   $y = \dfrac{1}{2}x + 1$    $m =$ _____ .   $b =$ _____    $y$-intercept: _____

   $y = -x + 4$    $m =$ _____   $b =$ _____    $y$-intercept: _____

   The lines intersect at point _____.

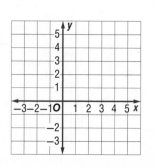

**2** Solve the system of equations by graphing.
$4x - 3y = 12$     $6y = 8x - 30$

**Step 1** Write each equation in _____.

$$4x - 3y = 12$$

$$-3y = \underline{\hspace{1cm}} + 12$$

_____

$$6y = 8x - 30$$

$$y = \frac{8}{6}x - \frac{30}{6}$$

_____

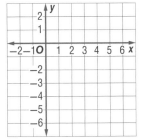

**Step 2** Graph both lines on the same grid.

**Step 3** The lines are _____.

**Solve each system of equations by graphing.**

**3** $y = \frac{2}{3}x + 3$
$x + y = -2$

_____

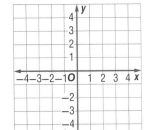

**4** $x - y = -8$
$y + x = 2$

_____

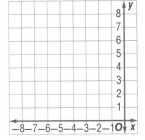

**5** NUMBER SENSE   The sum of two numbers is 5. The difference between the two numbers is 3. Find the two numbers.

$x + y = 5$

$y = $ _____

$x - y = 3$

$y = $ _____

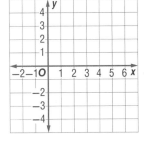

Check off each step.

_____ **Understand: I underlined key words.**

_____ **Plan: To solve the problem, I will** _____.

_____ **Solve: The answer is** _____.

_____ **Check: I checked my answer by** _____.

_____. GO ON

# ▶ Skills, Concepts, and Problem Solving

**Solve each system of equations by graphing.**

**6** $y = 2x$

$y + x = 3$

_____

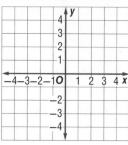

**7** $3y = -2x + 6$

$\frac{2}{3}x = -y - 1$

_____

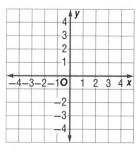

**8** $y - 2 = x$

$x = -\frac{1}{2}y - 2$

_____

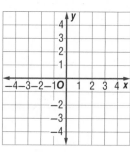

**9** $y + 3x = -1$

$2 + 2y = -6x$

_____

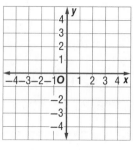

**Solve. Write the answer in simplest form.**

**10** **PERIMETER**  The length of a rectangle is 4 meters longer than its width. Use the equations $20 = 2\ell + 2w$ and $\ell = w + 4$ to find the length and width of the rectangle.

| perimeter = 20 m |
| --- |

_____

**11** **TEST**  A test worth 50 points has 10 questions on it. The multiple choice questions are worth 1 point and the essay questions are worth 6 points. How many of each type of questions are on the test?

_____

**Vocabulary Check**  **Write the vocabulary word that completes each sentence.**

**12** A function whose graph is a line is called a(n) _____.

**13** A(n) _____ is a set of two or more equations with the same variables.

**14** **Reflect**  Describe the graphs of a system of two linear equations with one solution, no solutions, and infinite solutions.

_____

_____

_____

# Solve Systems of Linear Equations Algebraically

## KEY Concept

There are two algebraic methods to solve **systems of equations**.

### Substitution Method

Solve one equation for a single variable. Substitute the expression into the second equation.

$$y = x + 2 \qquad\qquad y = 2x$$

$$2x = x + 2$$

$$\underline{-x \quad -x}$$

$$x = 2$$

> Substitute $2x$ for $y$.

$$y = 2x$$

$$y = 2(2)$$

$$y = 4$$

The solution is $(2, 4)$. Use this method when one equation is already solved for one of the variables.

### Elimination Method

Add or subtract multiples of the equations to eliminate a variable.

$$2x + 4y = 16 \rightarrow \qquad\qquad \rightarrow \quad 2x + 4y = 16$$

$$-3x + 2y = 0 \rightarrow -2(-3x + 2y = 0) \rightarrow + \underline{(6x - 4y = 0)}$$

$$8x = 16$$

$$x = 2$$

> The LCM of 4 and 2 is 4.

Substitute to find the value of $y$.

$$2x + 4y = 16$$

$$2(2) + 4y = 16$$

$$4 + 4y = 16$$

$$\underline{-4 \qquad\quad -4}$$

$$\frac{4y}{4} = \frac{12}{4}$$

$$y = 3$$

$$(2, 3)$$

Use this method when the coefficients of the variables are opposites or their **least common multiple** is recognizable.

## VOCABULARY

**least common multiple (LCM)**
the least of the common multiples of two or more numbers

**linear function**
a function with ordered pairs that satisfy a linear equation

**solution of a system of equations**
an ordered pair that satisfies both equations

**system of equations**
a set of equations with the same variables

**GO ON**

A solution to a system must satisfy both equations. Substitute the ordered pair into each equation to verify that it makes a true statement.

## Example 1

**Solve the system of equations by substitution.**

$$3x - y = -1$$
$$x = y + 1$$

1. The second equation is solved for $x$.

2. Substitute $y + 1$ for $x$ in the first equation. Solve for $y$.

| | |
|---|---|
| $3(y + 1) - y = -1$ | Substitute. |
| $3y + 3 - y = -1$ | Multiply. |
| $2y + 3 = -1$ | Combine like terms. |
| $2y = -4$ | Subtract. |
| $y = -2$ | Divide. |

3. Substitute the solution for $y$ into one of the equations. Solve for $x$.

$$x = y + 1$$
$$x = -2 + 1$$
$$x = -1$$

4. The solution is (-1, -2).

### YOUR TURN!

**Solve the system of equations by substitution.**

$$y = -x - 6$$
$$y - 3x = -2$$

1. The first equation is solved for _____.

2. Substitute _____ for _____ in the second equation. Solve for $x$.

_____ $- 3x = -2$ _____

_____ $= -2$ _____

_____ $= 4$ _____

$x =$ _____ _____

3. Substitute the solution for $x$ into one of the equations. Solve for $y$.

$$y = -(\underline{\phantom{xx}}) - 6$$
$$y = \underline{\phantom{xx}} - 6$$
$$y = \underline{\phantom{xx}}$$

4. The solution is (_____, _____).

## Example 2

**Solve the system of equations by elimination.**

$$4x + 2y = 4$$
$$6x + 2y = 8$$

*Eliminate y because the coefficients are the same.*

1. Subtract the equations to eliminate one variable. Distribute the negative.

$$\begin{array}{rcl} 4x + 2y = 4 & \rightarrow & 4x + 2y = 4 \\ -(6x + 2y = 8) & \rightarrow & \underline{-6x - 2y = -8} \\ & & -2x = -4 \end{array}$$

2. Solve for $x$.

$$\dfrac{-2x}{-2} = \dfrac{-4}{-2} \qquad \text{Divide by } -2.$$

$$x = 2$$

3. Substitute the solution into one of the original equations and solve.

$$4x + 2y = 4$$
$$4(2) + 2y = 4 \qquad \text{Substitute.}$$
$$8 + 2y = 4 \qquad \text{Multiply.}$$
$$2y = -4 \qquad \text{Subtract.}$$
$$y = -2 \qquad \text{Divide.}$$

4. The solution is $(2, -2)$.

---

**YOUR TURN**

**Solve the system of equations by elimination.**

$$x + 3y = 10$$
$$x + y = 6$$

1. _____ the equations to eliminate one variable. Distribute the negative.

$$\begin{array}{rcl} x + 2y = 10 & \rightarrow & x + 3y = 10 \\ -(x + y = 6) & \rightarrow & \underline{\quad} = \underline{\quad} \\ & & \underline{\quad} = \underline{\quad} \end{array}$$

2. Solve for $y$.

$$\underline{\quad} = \underline{\quad} \qquad \underline{\hspace{3cm}}$$
$$\underline{\quad} = \underline{\quad}$$

3. Substitute the solution into one of the original equations and solve.

$$x + y = 6$$
$$x + \underline{\quad} = 6 \qquad \underline{\hspace{3cm}}$$
$$x = \underline{\quad} \qquad \underline{\hspace{3cm}}$$

4. The solution is $(\underline{\quad}, \underline{\quad})$.

---

 **Guided Practice**

**Solve each system of equations by substitution.**

**1** $2x + 2y = 18 \qquad y = 2x$

$$2x + 2(\underline{\quad}) = 18 \qquad \text{Substitute.}$$
$$\underline{\hspace{3cm}} = \underline{\quad} \qquad \text{Multiply.}$$
$$\underline{\hspace{3cm}} = \underline{\quad} \qquad \text{Add.}$$
$$\underline{\hspace{3cm}} \qquad \text{Divide.}$$
$$\underline{\hspace{3cm}} = \underline{\quad}$$
$$y = 2(\underline{\quad}) = 6 \qquad (\underline{\quad}, \underline{\quad})$$

**2** $x + y = 4 \qquad y = x$

$$x + (\underline{\quad}) = 4$$
$$\underline{\hspace{3cm}} = \underline{\quad} \qquad \underline{\hspace{3cm}}$$
$$\underline{\hspace{3cm}} \qquad \underline{\hspace{3cm}}$$
$$\underline{\hspace{3cm}} = \underline{\quad}$$
$$(\underline{\quad}) + y = 4$$
$$y = 2 \qquad (\underline{\quad}, \underline{\quad})$$

## Step by Step Practice

3 Solve the system of equations by elimination.

$$4x + 4y = -4$$
$$-4x + 7y = 26$$

**Step 1** Add the equations to eliminate one variable.

$$-4x + 7y = 26$$

$$+ (\underline{\hspace{3cm}})$$

$$\underline{\hspace{3cm}}$$

$$y = \underline{\hspace{1cm}}$$

**Step 2** Substitute the solution into one of the original equations and solve.

$$4x + 4(\underline{\hspace{1cm}}) = -4 \qquad \text{Substitute.}$$

$$\underline{\hspace{4cm}} = \underline{\hspace{3cm}} \qquad \text{Multiply.}$$

$$\underline{\hspace{5cm}} \qquad \text{Subtract.}$$

$$\underline{\hspace{4cm}} = \underline{\hspace{3cm}} \qquad \text{Divide.}$$

$$x = \underline{\hspace{1cm}}$$

**Step 3** The solution is $(\underline{\hspace{1cm}}, \underline{\hspace{1cm}})$.

**Solve each system of equations by elimination.**

4  $x + y = 2$
   $x - y = 4$
   Add the two equations.

$$x + y = 2$$
$$+ (\underline{\hspace{2.5cm}})$$
$$\underline{\hspace{2.5cm}}$$
$$x = \underline{\hspace{1cm}}$$
$$\underline{\hspace{1cm}} + y = 2$$
$$y = \underline{\hspace{1cm}}$$

The solution is $(\underline{\hspace{1cm}}, \underline{\hspace{1cm}})$.

5  $4x + 3y = 27$
   $4x - 2y = 2$
   Subtract the two equations.

$$4x + 3y = 27 \quad \rightarrow \quad \underline{\hspace{3cm}}$$
$$- (\underline{\hspace{2.5cm}}) \rightarrow \underline{\hspace{3cm}}$$
$$\underline{\hspace{1cm}} = \underline{\hspace{1cm}}$$
$$y = \underline{\hspace{1cm}}$$
$$4x + 3(\underline{\hspace{1cm}}) = 27$$
$$4x + \underline{\hspace{1cm}} = 27$$
$$\underline{\hspace{1cm}} = \underline{\hspace{1cm}}$$
$$x = \underline{\hspace{1cm}}$$

The solution is $(\underline{\hspace{1cm}}, \underline{\hspace{1cm}})$.

**Solve.**

**6** Omar and Noah together weigh 104 pounds. Omar's weight, $x$, is 10 pounds less than twice Noah's weight, $y$. Find weights of Omar and Noah.

Solve by _____.          $x + y = 104$     $x = 2y - 10$

Omar: $x =$ _____

Noah: $y =$ _____

Check off each step.

_____ Understand: I underlined key words.

_____ Plan: To solve the problem, I will _____.

_____ Solve: The answer is _____.

_____ Check: I checked my answer by _____

_____.

 ## Skills, Concepts, and Problem Solving

**Solve each system of equations.**

**7**  $2x + y = 8$          (_____, _____)

$5x - y = 6$

**8**  $3x - 4y = 26$     (_____, _____)

$x + 2y = 2$

**9**  $-10x - 5y = 40$   (_____, _____)

$-2x + y = 0$

**10**  $2x + 3y = -49$   (_____, _____)

$x - 5y = 8$

GO ON

**Solve each system of equations.**

**11** $5x + 2y = 48$  (_____, _____)

   $3x + 2y = 32$

**12** $4x + y = 26$  (_____, _____)

   $y = x - 4$

**Solve.**

**13** **TICKET SALES**  East High School is hosting the championship basketball game. The ticket prices are shown on the sign. If 1,000 tickets were sold for $3,110, how many of each type of ticket were sold?

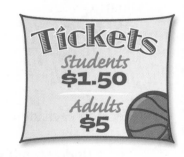

**14** **MONEY**  Krista has 5 times as much money as Grace does. Together, Krista and Grace have $312. How much money does each girl have?

**Vocabulary Check**  **Write the vocabulary word that completes each sentence.**

**15** Two ways to solve a system of equations algebraically are

   elimination and _____.

**16** A system of equations is a set of equations that have _____.

**17** A linear function has ordered pairs that satisfy a _____.

**18** **Reflect**  Describe either the substitution or elimination method to solve a system of equations. Give an example. Tell why you would use the method you described.

_____

_____

_____

_____

STOP

**Solve each system by graphing.**

**1** $y = x + 7$

$y = -2x - 2$

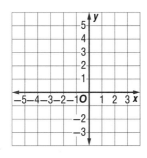

**2** $y = -x - 2$

$y = 3x - 6$

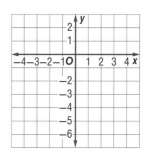

**Solve each system of equations.**

**3** $y - 2x = -5$ _____

$y + 2x = 11$

**4** $y - 3x = 10$ _____

$y - x = 6$

**5** $-2x + y = -1$ _____

$-4x + 2y = 6$

**6** $y - \frac{1}{2}x = -3$ _____

$2y - x = -6$

**Solve.**

**7** **PETS** The equation that models the fees for Doggy Daycare is $y = 25x + 250$, where $x$ is the number of days a dog is boarded. The equation that models the fees for Pretty Puppy is $y = 15x + 290$, where $x$ is the number of days a dog is boarded. At what number of days will the charges be the same for the two dog care facilities? How much will the charges be?

**8** **ELECTIONS** In a local election, the number of votes for the Party A candidate was 4 times higher than for the Party B candidate. The total number of votes was 1,435. Use the equations $a = 4b$ and $a + b = 1,435$ to find how many votes were for the Party A candidate.

# Chapter Test

**Identify the domain and range of each relation. Tell whether the relation is a function.**

**1**
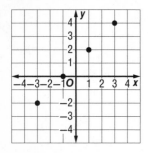

domain = _____

range = _____

The relation _____ a function.

**2**
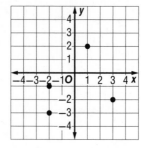

domain = _____

range = _____

The relation _____ a function.

**3**

| x | y = −2x − 5 | y |
|---|---|---|
| 1 | y = −2(1) − 5 | −7 |
| 2 | y = −2(2) − 5 | −9 |
| 3 | y = −2(3) − 5 | −11 |

domain = _____

range = _____

The relation _____ a function.

**4** {(−1, 2), (0, 3), (−1, 3), (4, 6)}

domain = _____

range = _____

The relation _____ a function.

**Find the slope of each line.**

**5**
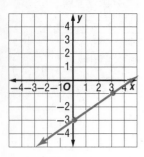

_____

**6**
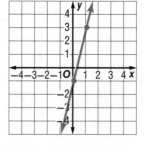

_____

**Find the slope of the line that contains each pair of points.**

**7** (−2, 1) and (4, 3)

**8** (6, −1) and (2, −2)

**Write the equation of each line in slope-intercept form.**

**9** $3y + 6x = 12$

**10** $y + 1 = \dfrac{1}{2}x$

**Write the equation for the line that passes through each given point and has each given slope.**

**11** $(1, 1)$, $m = 3$

**12** $(0, -3)$, $m = \dfrac{1}{2}$

**Graph each equation.**

**13** $y = -\dfrac{3}{4}x - 3$

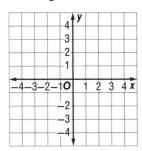

**14** $2y + 2x = 8$

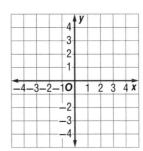

**Solve each system of equations.**

**15** $y = -x + 3$
$y = -3x + 11$

**16** $2x - y = 15$
$x - 2y = 12$

**17** $2x + 3y = 6$
$x + \dfrac{3}{2}y = 3$

**Solve. Write the answer in simplest terms.**

**18** **ACCESSIBILITY** The wheelchair access ramp at Fannie's school is created as shown below. What is the slope of the ramp?

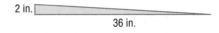

2 in.
36 in.

_____

**Correct the mistake.**

**19** **HOME IMPROVEMENT** Plumber A charges $30 for a visit, plus $25 per hour for labor Plumber B charges only an hourly rate of $35. Mr. Wilson told his wife that she could call either plumber because the fee would be the same. Is Mr. Wilson correct?

_____

_____

STOP

# Chapter 5

# Measurement

## Are you tall enough?

To ride a roller coaster, you must be 48 inches tall. If a person is 4.2 feet tall, is that tall enough? To find out, you can convert feet to inches.

**STEP 2 Preview** Get ready for Chapter 5. Review these skills and compare them with what you will learn in this chapter.

| What You Know | What You Will Learn |
|---|---|
| You know how to multiply by powers of ten. | *Lesson 5-4* |

**What You Know**

You know how to multiply by powers of ten.

**Example:**
$3.6 \cdot 10 = 36$

> To multiply $3.6 \cdot 10$, move the decimal point 1 place to the right.

**TRY IT!**

1  $7.2 \cdot 10 =$ _____

2  $8.21 \cdot 100 =$ _____

3  $0.624 \cdot 1,000 =$ _____

4  $2.34 \cdot 1,000 =$ _____

**What You Will Learn**

*Lesson 5-4*

To change a metric measure from one unit to another, you can multiply or divide by a power of 10.

Convert 3.75 meters to millimeters.

Use the relationship 1 m = 1,000 mm.

$3.75 \cdot 1,000 = 3750$

So, 3.75 m = 3750 mm.

---

**What You Know**

You know how to simplify expressions.

**Example:** $2(8) + 2(11) = 16 + 22$
$= 38$

**TRY IT!**

1  $4(22) =$ _____

2  $4(18) =$ _____

3  $2(7) + 2(14) =$ _____

4  $2(8) + 2(15) =$ _____

**What You Will Learn**

*Lesson 5-7*

The perimeter $P$ of a rectangle is the sum of the lengths and widths. It is also two times the length $\ell$ plus two times the width $w$.

7 ft, 15 ft

$P = 2\ell + 2w$
$P = 2(7) + 2(15)$
$P = 14 + 30$
$P = 44$

The perimeter is 44 feet.

# Length in the Customary System

## KEY Concept

The **customary system** is used throughout the United States with units of **length** such as inches, feet, yards, and miles.

| Unit | Abbreviation | Equivalent | Example |
|------|-------------|-----------|---------|
| inch | in. | — | paperclip |
| foot | ft | 1 ft = 12 in. | notebook |
| yard | yd | 1 yd = 3 ft | baseball bat |
| mile | mi | 1 mi = 1,760 yd | 15 football fields |

To convert between units, write **proportions** so that units can cancel.

Cancel units.

$$\frac{24 \ \cancel{ft}}{1} \cdot \frac{1 \ yd}{3 \ \cancel{ft}}$$

Cancel factors.

$$\frac{\overset{8}{\cancel{24}}}{1} \cdot \frac{1 \ yd}{\underset{1}{\cancel{3}}} = \frac{8 \ yd}{1} = 8 \ yd$$

Divide 24 and 3 by 3 to reduce the fraction.

### VOCABULARY

**customary system**
a measurement system that includes units such as foot, pound, and quart

**length**
a measurement of the distance between two points

**proportion**
an equation of the form $\frac{a}{b} = \frac{c}{d}$ stating that two ratios are equivalent

Some conversions may require two proportions.

## Example 1

**Convert.  7 yd = _____ ft**

1. Set up a proportion so that units cancel.

$$\frac{7 \ yd}{1} \cdot \frac{3 \ ft}{1 \ yd}$$

2. Multiply and simplify.

$$\frac{7}{1} \cdot \frac{3 \ ft}{1} = 21 \ ft$$

7 yd = 21 ft

### YOUR TURN!

**Convert.  5,280 yd = _____ mi**

1. Set up a proportion so that units cancel.

2. Multiply and simplify.

5,280 yd = _____

## Example 2

**Convert.   120 in. = _____ yd**

1. Set up a proportion so that units cancel.

Cancel in. and ft.

$$\frac{120 \text{ in.}}{1} \cdot \frac{1 \text{ ft}}{12 \text{ in.}} \cdot \frac{1 \text{ yd}}{3 \text{ ft}}$$

2. Multiply and simplify.

$$\frac{\overset{10}{\cancel{120}}}{1} \cdot \frac{1}{\underset{1}{\cancel{12}}} \cdot \frac{1 \text{ yd}}{3} = \frac{10 \text{ yd}}{3}$$

Divide 120 and 12 by 12.

$$120 \text{ in.} = 3\frac{1}{3} \text{ yd}$$

---

**YOUR TURN!**

**Convert.   5.3 mi = _____ ft**

1. Set up a proportion so that units cancel.

$$\frac{\boxed{\phantom{xx}}}{1} \cdot \frac{\boxed{\phantom{xx}}}{\boxed{\phantom{xx}}} \cdot \frac{\boxed{\phantom{xx}}}{\boxed{\phantom{xx}}}$$

2. Multiply and simplify.

$$\frac{\boxed{\phantom{xx}}}{1} \cdot \frac{\boxed{\phantom{xx}}}{\boxed{\phantom{xx}}} \cdot \frac{\boxed{\phantom{xx}}}{\boxed{\phantom{xx}}} = \frac{\boxed{\phantom{xx}}}{\boxed{\phantom{xx}}}$$

5.3 mi = _____

---

## ▶ Guided Practice

**Convert each measurement.**

**1**  11 yd = _____ ft

$$\frac{\boxed{\phantom{xx}}}{1} \cdot \frac{3 \text{ ft}}{1 \text{ yd}} = \text{_____} = \text{\_\_\_\_\_}$$

**2**  2 mi = _____ yd

$$\frac{\boxed{\phantom{xx}}}{1} \cdot \frac{1,760 \text{ yd}}{1 \text{ mi}} = \text{_____} = \text{\_\_\_\_\_}$$

**3**  56 in. = _____ ft

$$\frac{56 \text{ in.}}{1} \cdot \frac{\boxed{\phantom{xx}}}{12 \text{ in.}} = \frac{\boxed{\phantom{xx}}}{\boxed{\phantom{xx}}} = \text{_____}$$

**4**  15 ft = _____ yd

$$\frac{15 \text{ ft}}{1} \cdot \frac{\boxed{\phantom{xx}}}{3 \text{ ft}} = \frac{\boxed{\phantom{xx}}}{\boxed{\phantom{xx}}} = \text{\_\_\_\_\_}$$

**5**  9 ft = _____ in.

$$\frac{\boxed{\phantom{xx}}}{1} \cdot \frac{\boxed{\phantom{xx}}}{\boxed{\phantom{xx}}} = \text{_____} = \text{\_\_\_\_\_}$$

**6**  8,800 yd = _____ mi

$$\frac{\boxed{\phantom{xx}}}{1} \cdot \frac{\boxed{\phantom{xx}}}{\boxed{\phantom{xx}}} = \frac{\boxed{\phantom{xx}}}{\boxed{\phantom{xx}}} = \text{\_\_\_\_\_}$$

**GO ON** ➡

7 Convert. 126,720 in. = ____ mi

**Step 1** Set up a proportion so that units cancel.

[ 5,280 ft = 1 mi ]

$$\frac{126{,}720 \text{ in.}}{1} \cdot \boxed{\phantom{xxxx}} \cdot \boxed{\phantom{xxxx}}$$

**Step 2** Multiply and simplify.

$$\frac{\phantom{xxxxxxxx} \cdot \phantom{xxxxxx} \cdot \phantom{xxxxxx}}{}$$

$$= \frac{\phantom{xxxxxxx}}{} = \underline{\phantom{xxx}}$$

**Step 3** 126,720 in. = _____

---

**Convert each measurement.**

8 216 in. = ____ yd

$$\frac{216 \text{ in.}}{1} \cdot \frac{1 \text{ ft}}{12 \text{ in.}} \cdot \frac{1 \text{ yd}}{3 \text{ ft}} = \frac{\boxed{\phantom{xx}}}{\boxed{\phantom{xx}}} = \underline{\phantom{xxxxx}}$$

9 15 yd = ____ in.

$$\frac{\boxed{\phantom{x}} \text{ yd}}{1} \cdot \frac{\boxed{\phantom{x}} \text{ ft}}{1 \text{ yd}} \cdot \frac{\boxed{\phantom{x}} \text{ in.}}{1 \text{ ft}} = \underline{\phantom{xxxxxxxx}} = \underline{\phantom{xxxx}}$$

10 7 mi = ____ ft

$$\frac{\boxed{\phantom{xx}}}{1} \cdot \frac{\boxed{\phantom{xx}}}{\boxed{\phantom{xx}}} \cdot \frac{\boxed{\phantom{xx}}}{\boxed{\phantom{xx}}} = \underline{\phantom{xxxxxxx}} = \underline{\phantom{xxxxx}}$$

11 126 in. = ____ yd

$$\frac{\boxed{\phantom{xx}}}{\boxed{\phantom{xx}}} \cdot \frac{\boxed{\phantom{xx}}}{\boxed{\phantom{xx}}} \cdot \frac{\boxed{\phantom{xx}}}{\boxed{\phantom{xx}}} = \underline{\phantom{xxxx}} = \underline{\phantom{xxxxx}}$$

**Solve.**

12  FENCING   Samir is putting a fence around his backyard.
The dimensions of his rectangular backyard are 38 feet by
20 feet. The fencing is sold in 1-yard sections. How many
sections of fence does he need to buy?

The distance around the yard is

_____ + _____ + _____ + _____ = _____.

Set up a proportion to find the number of yards.

Check off each step.

_____ **Understand: I underlined key words.**

_____ **Plan: To solve the problem, I will** _____.

_____ **Solve: The answer is** _____.

_____ **Check: I checked my answer by** _____

_____.

## Skills, Concepts, and Problem Solving

**Convert each measurement.**

13  6.5 ft = _____ in.

$$\frac{\boxed{\phantom{x}}}{1} \cdot \frac{\boxed{\phantom{x}}}{\boxed{\phantom{x}}} = \text{_____} = \text{\_\_\_\_}$$

14  198 in. = _____ yd

$$\frac{\boxed{\phantom{x}}}{1} \cdot \frac{\boxed{\phantom{x}}}{\boxed{\phantom{x}}} \cdot \frac{\boxed{\phantom{x}}}{\boxed{\phantom{x}}} = \frac{\boxed{\phantom{x}}}{\boxed{\phantom{x}}} = \text{\_\_\_\_}$$

15  2,640 ft = _____ mi

_____ = _____

16  8 yd = _____ ft

_____ = _____

GO ON

**Convert each measurement.**

**17** 7.5 ft = _____ yd

**18** 504 in. = _____ yd

**19** 3,520 yd = _____ mi

**20** 3 mi = _____ in.

**Solve.**

**21** **MEASUREMENT** Pedro is 66 inches tall. How tall is Pedro in feet?

_____

**22** **BASEBALL** Taro knows that the distance between the bases on the softball diamond is 90 feet. Taro just hit a homerun and wants to know how many inches she ran.

_____

**23** **KNITTING** Susan is knitting a blanket. She needs 1,512 inches of blue yarn. The yarn is only sold by the yard. How much blue yarn should she buy?

_____

**Vocabulary Check** **Write the vocabulary word that completes each sentence.**

**24** The measurement system that includes inches, feet, yards, and miles is the

_____.

**25** _____ are equations in the form of $\frac{a}{b} = \frac{c}{d}$ which state that two ratios are equivalent.

**26** **Reflect** Explain how to use proportions to convert customary units of length.

_____

_____

_____

STOP

# Capacity in the Customary System

Copyright © Glencoe/McGraw-Hill, a division of The McGraw-Hill Companies, Inc.

## KEY Concept

**Capacity** is the amount of dry or liquid material that a container holds.

| Unit | Abbreviation | Equivalent | Example |
|------|-------------|-----------|---------|
| fluid ounce | fl oz | — | eye dropper |
| cup | c | 1 c = 8 fl oz | coffee mug |
| pint | pt | 1 pt = 2 c | cereal bowl |
| quart | qt | 1 qt = 2 pt | pitcher |
| gallon | gal | 1 gal = 4 qt | milk carton |

To convert between units, write **proportions** so that units can cancel.

Cancel units.

$$\frac{24 \; \cancel{pt}}{1} \cdot \frac{1 \; qt}{2 \; \cancel{pt}}$$

Cancel factors.

$$\frac{\overset{12}{\cancel{24}}}{1} \cdot \frac{1 \; qt}{\underset{1}{\cancel{2}}} = \frac{12 \; qt}{1} = 12 \; qt$$

### VOCABULARY

**capacity**
the amount of dry or liquid material a container can hold

**customary system**
a measurement system that includes units such as foot, pound, and quart

**proportion**
an equation of the form $\frac{a}{b} = \frac{c}{d}$ stating that two ratios are equivalent

Nearly all food products are sold based on the capacity, or volume, of the product.

## Example 1

**Convert.   64 pt = _____ qt**

1. Set up a proportion so that units cancel.

$$\frac{64 \; \cancel{pt}}{1} \cdot \frac{1 \; qt}{2 \; \cancel{pt}}$$

Cancel pt.

2. Multiply and simplify.

$$\frac{\overset{32}{\cancel{64}}}{1} \cdot \frac{1 \; qt}{\underset{1}{\cancel{2}}}$$

Divide 64 and 2 by 2.

64 pt = 32 qt

## YOUR TURN!

**Convert.   2 c = _____ fl oz**

1. Set up a proportion so that units cancel.

2. Multiply and simplify.

2 c = _____

GO ON

## Example 2

**Convert.   48 c = _____ gal**

1. Set up a proportion so that units cancel.

$$\frac{48\ c}{1} \cdot \frac{1\ pt}{2\ c} \cdot \frac{1\ qt}{2\ pt} \cdot \frac{1\ gal}{4\ qt}$$

2. Multiply and simplify.

$$\frac{\overset{\overset{3}{\cancel{12}}}{\cancel{48}}}{1} \cdot \frac{1}{\underset{1}{\cancel{2}}} \cdot \frac{1}{\underset{1}{\cancel{2}}} \cdot \frac{1\ gal}{\underset{1}{\cancel{4}}} = \frac{3\ gal}{1}$$

48 c = 3 gal

---

**Convert.   22.5 gal = _____ pt**

1. Set up a proportion so that units cancel.

2. Multiply and simplify.

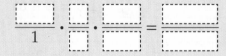

22.5 gal = _____

---

## ▶ Guided Practice

**Convert each measurement.**

**1**   24 qt = _____ gal

$$\frac{\boxed{\phantom{xx}}\ qt}{\boxed{\phantom{x}}} \cdot \frac{1\ gal}{4\ qt} = \underline{\hspace{2cm}} = \underline{\hspace{2cm}}$$

**2**   96 fl oz = _____ c

$$\frac{\boxed{\phantom{xxxx}}}{1} \cdot \frac{1\ c}{8\ fl\ oz} = \underline{\hspace{1.5cm}} = \underline{\hspace{1.5cm}}$$

**3**   4.5 gal = _____ qt

$$\cdot = \underline{\hspace{1.5cm}}$$

**4**   11 pt = _____ c

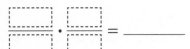

$$\cdot = \underline{\hspace{1.5cm}}$$

## Step by Step Practice

**5** Convert.   12 c = _____ qt

**Step 1**   Set up a proportion so that units cancel.

$$\frac{12\text{ c}}{1} \cdot \frac{\boxed{\phantom{xx}}}{\boxed{\phantom{xx}}} \cdot \frac{\boxed{\phantom{xx}}}{\boxed{\phantom{xx}}}$$

**Step 2**   Multiply and simplify.

$$\underline{\hspace{2cm}} \cdot \underline{\hspace{2cm}} \cdot \underline{\hspace{2cm}} = \underline{\hspace{2cm}}$$

**Step 3**   12 c = _____

---

**6**   1 gal = _____ c

$$\frac{1\ \cancel{\text{gal}}}{1} \cdot \frac{\boxed{\phantom{xx}}}{\boxed{\phantom{xx}}} \cdot \frac{\boxed{\phantom{xx}}}{\boxed{\phantom{xx}}} \cdot \frac{2\text{ c}}{1\ \cancel{\text{pt}}} = \underline{\hspace{2.5cm}}$$

**7**   50 c = _____ qt

$$\frac{50\ \cancel{\text{c}}}{1} \cdot \frac{\boxed{\phantom{xx}}}{\boxed{\phantom{xx}}} \cdot \frac{\boxed{\phantom{xx}}}{\boxed{\phantom{xx}}} = \underline{\hspace{2.5cm}}$$

**8**   3 pt = _____ fl oz

$$\frac{\boxed{\phantom{xx}}}{\boxed{\phantom{xx}}} \cdot \frac{\boxed{\phantom{xx}}}{\boxed{\phantom{xx}}} \cdot \frac{\boxed{\phantom{xx}}}{\boxed{\phantom{xx}}} = \underline{\hspace{2.5cm}}$$

**9**   5 qt = _____ fl oz

$$\frac{\boxed{\phantom{xx}}}{\boxed{\phantom{xx}}} \cdot \frac{\boxed{\phantom{xx}}}{\boxed{\phantom{xx}}} \cdot \frac{\boxed{\phantom{xx}}}{\boxed{\phantom{xx}}} \cdot \frac{\boxed{\phantom{xx}}}{\boxed{\phantom{xx}}} = \underline{\hspace{2.5cm}}$$

**10**   12 qt = _____ c

$$\frac{\boxed{\phantom{xx}}}{\boxed{\phantom{xx}}} \cdot \frac{\boxed{\phantom{xx}}}{\boxed{\phantom{xx}}} \cdot \frac{\boxed{\phantom{xx}}}{\boxed{\phantom{xx}}} = \underline{\hspace{2.5cm}}$$

GO ON

## Step by Step Problem-Solving Practice

**Solve.**

**11** **PRINTING** Robert knows that he used 280 fluid ounces of ink to print posters for an exhibition. How many pints of ink did Robert use?

$$\frac{280 \text{ fl oz}}{1} \cdot \frac{\boxed{\phantom{xx}}}{\boxed{\phantom{xx}}} \cdot \frac{\boxed{\phantom{x}}}{\boxed{\phantom{x}}} =$$

$$\frac{\boxed{\phantom{xxxx}}}{1} \cdot \frac{\boxed{\phantom{xx}}}{\boxed{\phantom{xx}}} \cdot \frac{\boxed{\phantom{xx}}}{\boxed{\phantom{xx}}} = \frac{\boxed{\phantom{x}}}{1} \cdot \frac{\boxed{\phantom{x}}}{\boxed{\phantom{x}}} \cdot \frac{\boxed{\phantom{x}}}{\boxed{\phantom{x}}}$$

$$= \underline{\hspace{2cm}}$$

$$= \underline{\hspace{2cm}}$$

Check off each step.

_____ Understand: I underlined key words.

_____ Plan: To solve the problem, I will _____.

_____ Solve: The answer is _____.

_____ Check: I checked my answer by _____.

 Skills, Concepts, and Problem Solving

**Convert each measurement.**

**12** 36 p = _____ qt

_____ = _____

**13** 9 c = _____ fl oz

_____ = _____

**14** 2.5 gal = _____ fl oz

**15** 112 fl oz = _____ pt

**16** 144 pt = _____ gal

**17** 15.5 qt = _____ c

**18** 4 gal = _____ c

**19** 20 c = _____ qt

**156** **Chapter 5** Measurement

Copyright © Glencoe/McGraw-Hill, a division of The McGraw-Hill Companies, Inc.

**Solve.**

**20** SCHOOL LUNCH   The cafeteria sells 228 pints of milk on an average lunch day. How many cups of milk does the cafeteria sell on an average day?

_____

**21** GROCERY   Ling sees a sale on a 5-gallon container of water at the store. How many cups of water are in the container?

_____

**22** WATER CONSERVATION   Anna is trying to conserve water at home. She uses 25 gallons of water each time she showers. With a new shower head she will only use 20 gallons for each shower. How many quarts of water will she save during one week if she takes one shower each day?

_____

**Vocabulary Check**   **Write the vocabulary word that completes each sentence.**

**23** The amount of dry or liquid material a container can hold is called

the _____ of the container.

**24** Cups, pints, and quarts are all units of measure used in the

_____ system of measurement.

**25** Reflect   Write a proportion using ounce, cup, and pint that is not one of the conversions given. Explain why the proportion could be helpful.

_____

_____

_____

# Progress Check 1 (Lessons 5-1 and 5-2)

## Convert each measurement.

1  3,520 yd = _____ mi

2  40 ft = _____ in.

3  108 in. = _____ yd

4  13,200 ft = _____ mi

5  1 mi = _____ yd

6  648 in. = _____ yd

7  4 gal = _____ c

8  560 fl oz = _____ pt

9  6 qt = _____ fl oz

10  18 qt = _____ pt

11  10 gal = _____ qt

12  7 pt = _____ fl oz

## Solve.

13  **LAW ENFORCEMENT**   By law, you must stop at least 50 feet from railroad tracks. Kamilah received a ticket that stated she was stopped 550 inches from railroad tracks. Did Kamilah break the law? Explain.

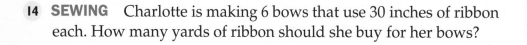

14  **SEWING**   Charlotte is making 6 bows that use 30 inches of ribbon each. How many yards of ribbon should she buy for her bows?

_____

15  **PARTY PLANNING**   Guests at a dinner party can be expected to drink at least 8 ounces of water with their meals. How many gallons of water should be provided for 35 guests?

_____

# Lesson 5-3

## Weight in the Customary System

### KEY Concept

The **customary system** has fewer units of measure for **weight**. Familiarize yourself with the following information.

| Unit | Abbreviation | Equivalent | Example |
|------|--------------|------------|---------|
| ounce | oz | — | one strawberry |
| pound | lb | 1 lb = 16 oz | loaf of bread |
| ton | T | 1 T = 2,000 lb | car |

To convert between units, write **proportions** so that units can cancel.

Cancel units.
$$\frac{48 \cancel{\text{ oz}}}{1} \cdot \frac{1 \text{ lb}}{16 \cancel{\text{ oz}}}$$

Cancel factors.
$$\frac{\overset{3}{\cancel{48}}}{1} \cdot \frac{1 \text{ lb}}{\underset{1}{\cancel{16}}} = \frac{3 \text{ lb}}{1} = 3 \text{ lb}$$

Divide 48 and 16 by 16 to reduce the fraction.

### VOCABULARY

**customary system**
a measurement system that includes units such as foot, pound, and quart

**proportion**
an equation of the form $\frac{a}{b} = \frac{c}{d}$ stating that two ratios are equivalent

**weight**
a measurement that tells how heavy or light an object is

---

### Example 1

Convert.  3 T = _____ lb

1. Set up a proportion so that units cancel.

$$\frac{3 \cancel{\text{ T}}}{1} \cdot \frac{2,000 \text{ lb}}{1 \cancel{\text{ T}}}$$

2. Multiply and simplify.

$$\frac{3}{1} \cdot \frac{2,000 \text{ lb}}{1} = 6,000 \text{ lb}$$

3 T = 6,000 lb

### YOUR TURN!

Convert.  72 oz = _____ lb

1. Set up a proportion so that units cancel.

$$\frac{\boxed{\phantom{xx}}}{1} \cdot \frac{\boxed{\phantom{xx}}}{\boxed{\phantom{xx}}}$$

2. Multiply and simplify.

$$\frac{\boxed{\phantom{xx}}}{1} \cdot \frac{1 \text{ lb}}{\boxed{\phantom{xx}}} = \frac{\boxed{\phantom{xx}}}{\boxed{\phantom{xx}}} = \underline{\phantom{xxxx}}$$

72 oz = _____

GO ON

## Example 2

**Convert.** 1 T = _____ oz

1. Set up a proportion so that units cancel.

$$\frac{1\ \cancel{T}}{1} \cdot \frac{2,000\ \cancel{lb}}{1\ \cancel{T}} \cdot \frac{16\ oz}{1\ \cancel{lb}}$$

2. Multiply and simplify.

$$\frac{1}{1} \cdot \frac{2,000}{1} \cdot \frac{16\ oz}{1} = 32,000\ oz$$

1 T = 32,000 oz

### YOUR TURN!

**Convert.** 80,000 oz = _____ T

1. Set up a proportion so that units cancel.

$$\frac{\boxed{\phantom{xxx}}}{1} \cdot \frac{\boxed{\phantom{xxx}}}{\boxed{\phantom{xxx}}} \cdot \frac{\boxed{\phantom{x}}}{\boxed{\phantom{x}}}$$

2. Multiply and simplify.

$$\frac{\boxed{\phantom{xxx}}}{1} \cdot \frac{\boxed{\phantom{x}}}{\boxed{\phantom{x}}} \cdot \frac{\boxed{\phantom{x}}}{\boxed{\phantom{x}}} = \frac{\boxed{\phantom{x}}}{\boxed{\phantom{x}}} = \underline{\phantom{xxx}}$$

80,000 oz = _____

## Guided Practice

**Convert each measurement.**

1  2 lb = _____ oz

$$\frac{\boxed{\phantom{x}}\ lb}{1} \cdot \frac{\boxed{\phantom{x}}\ oz}{1\ lb} = \underline{\phantom{xxxxx}}$$

2  5.5 T = _____ lb

$$\frac{\boxed{\phantom{x}}\ T}{1} \cdot \frac{\boxed{\phantom{xx}}\ lb}{1\ T} = \underline{\phantom{xxxxx}}$$

3  24 oz = _____ lb

$$\frac{\boxed{\phantom{xx}}}{1} \cdot \frac{\boxed{\phantom{x}}}{\boxed{\phantom{x}}} = \underline{\phantom{xxxxx}}$$

4  96,000 oz = _____ lb

$$\frac{\boxed{\phantom{xx}}}{1} \cdot \frac{\boxed{\phantom{x}}}{\boxed{\phantom{x}}} = \underline{\phantom{xxxxx}}$$

## Step by Step Practice

5  Convert.   48,000 oz = _____ T

**Step 1**  Set up a proportion so that units cancel.

$$\frac{48,000\ oz}{1} \cdot \frac{\boxed{\phantom{x}}\ lb}{\boxed{\phantom{x}}\ oz} \cdot \frac{\boxed{\phantom{xx}}\ T}{\boxed{\phantom{xx}}\ lb}$$

**Step 2**  Multiply and simplify.

$$\underline{\phantom{xxxxxxxxxxxxxxxxxxxxxxxx}} = \underline{\phantom{xxxx}}$$

**Step 3**  48,000 oz = _____ T

**Convert each measurement.**

**6** 8.5 T = _____ lb

$$\frac{\boxed{\phantom{xx}}}{1} \cdot \frac{\boxed{\phantom{xxxx}}}{\boxed{\phantom{xxxx}}} = \frac{\boxed{\phantom{xx}}}{1} \cdot \frac{\boxed{\phantom{xxxx}}}{\boxed{\phantom{xxxx}}} = \underline{\phantom{xxxxx}} = \underline{\phantom{xxxx}}$$

**7** 400,000 oz = _____ T

$$\frac{\boxed{\phantom{xxxx}}}{1} \cdot \frac{\boxed{\phantom{xx}}}{\boxed{\phantom{xx}}} \cdot \frac{\boxed{\phantom{xxx}}}{\boxed{\phantom{xxx}}} = \frac{\boxed{\phantom{xxxx}}}{1} \cdot \frac{\boxed{\phantom{xx}}}{\boxed{\phantom{xx}}} \cdot \frac{\boxed{\phantom{xxx}}}{\boxed{\phantom{xxx}}} = \underline{\phantom{xxx}}$$

**8** 165 lb = _____ oz

$$\frac{\boxed{\phantom{x}}}{\boxed{\phantom{x}}} \cdot \frac{\boxed{\phantom{x}}}{\boxed{\phantom{x}}} = \underline{\phantom{xxxx}}$$

**9** 5 T = _____ oz

$$\frac{\boxed{\phantom{x}}}{\boxed{\phantom{x}}} \cdot \frac{\boxed{\phantom{xx}}}{\boxed{\phantom{xx}}} \cdot \frac{\boxed{\phantom{x}}}{\boxed{\phantom{x}}} = \underline{\phantom{xxxxx}}$$

## Step by Step Problem-Solving Practice

**Solve.**

**10** **HEALTH** Mrs. Beal knows that her baby, Samantha, weighs 168 ounces. How many pounds does Samantha weigh?

$$\frac{168 \text{ oz}}{1} \cdot \frac{\boxed{\phantom{xx}}}{\boxed{\phantom{xx}}} = \underline{\phantom{xxxxx}} = \underline{\phantom{xxx}} = \underline{\phantom{xxx}}$$

Check off each step.

_____ **Understand: I underlined key words.**

_____ **Plan: To solve the problem, I will** _____.

_____ **Solve: The answer is** _____.

_____ **Check: I checked my answer by** _____.

GO ON

# ▶ Skills, Concepts, and Problem Solving

**Convert each measurement.**

**11** 64 oz = _____ lb

**12** 4 T = _____ oz

**13** 19,000 lb = _____ T

**14** 75 lb = _____ oz

**15** 800,000 oz = _____ T

**16** 14 T = _____ lb

**Solve.**

**17** SHOPPING  Brooke needs a 2-pound can of pumpkin for a pie dessert. The cans of pumpkin are sold in ounces. How many ounces should be in the can she buys?

_____

**18** TRAVEL  Luis is checking his suitcase to get on a plane. The weight limit for a suitcase is 800 ounces. What is the weight limit in pounds?

_____

**19** RESTAURANT  An Italian restaurant chef orders tomato sauce in a large can as shown. How many pounds is the can of tomato sauce?

_____

Tomato Sauce
152 oz

**Vocabulary Check  Write the vocabulary word that completes each sentence.**

**20** The _____ for measuring weight includes the units of ounces, pounds, and tons.

**21** The _____ of an object tells how heavy or light the object is.

**22** Reflect  Marco says that 192 pounds equals 12 ounces. Is his answer reasonable? Explain why or why not.

_____

_____

STOP

# Length in the Metric System

The **metric system** is used throughout the world. It is based on powers of ten. The most commonly used units are millimeters, centimeters, meters, and kilometers. The base unit of **length** in the metric system is the **meter**.

| Unit | Abbreviation | Number of Meters |
|---|---|---|
| millimeter | mm | 0.001 m |
| centimeter | cm | 0.01 m |
| decimeter | dm | 0.1 m |
| **meter** | **m** | **1 m** |
| dekameter | dkm | 10 m |
| hectometer | hm | 100 m |
| kilometer | km | 1,000 m |

To convert between units, multiply or divide by powers of ten. When converting to smaller units, **multiply**. When converting to larger units, **divide**.

A centimeter is one unit smaller than a decimeter.

**5.0 dm = __ cm**

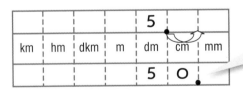

1 dm = 10 cm

Multiplying by 10 moves the decimal point 1 place to the right.

5 • 10 = 50 cm

A kilometer is two units larger than a dekameter.

**5.0 dkm = __ km**

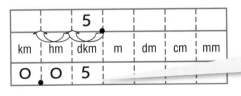

1 km = 100 dkm

Dividing by 100 moves the decimal point 2 places to the left.

$5 \div 10^2 = 5 \div 100 = 0.05$ km

VOCABULARY

**metric system**
a measurement system based on powers of 10 that includes units such as meter, gram, and liter

**meter**
a base unit in the metric system for measuring length

**length**
a measurement of the distance between two points

**GO ON**

## Example 1

**Convert.   8.2 m = _____ mm**

1. Are you converting to a smaller unit or a larger unit?

   smaller

2. Do you multiply or divide?

   multiply

3. How many units are you converting?

   three

4. Use $10^3$.

   $8.2 \cdot 10^3 =$

   $8.2 \cdot 1{,}000 = 8{,}200$

   $8.2 \text{ m} = 8{,}200 \text{ mm}$

## YOUR TURN!

**Convert.   1.3 km = _____ cm**

1. Are you converting to a smaller unit or a larger unit?

   _____

2. Do you multiply or divide?

   _____

3. How many units are you converting?

   _____

4. Use _____.

   $1.3 \underline{\phantom{xx}} 10^{\square} =$

   $1.3 \underline{\phantom{xxxxx}} = \underline{\phantom{xxxxx}}$

   $1.3 \text{ km} = \underline{\phantom{xxxxxxx}}$

## Example 2

**Convert.   613.5 mm =_____ dkm**

1. Are you converting to a smaller unit or a larger unit?

   larger

2. Do you multiply or divide?

   divide

3. How many units are you converting?

   four

4. Use $10^4$.

   $613.5 \div 10^4 =$

   $613.5 \div 10{,}000 = 0.06135$

   $613.5 \text{ mm} = 0.06135 \text{ dkm}$

## YOUR TURN!

**Convert.   2,234 m = _____ km**

1. Are you converting to a smaller unit or a larger unit?

   _____

2. Do you multiply or divide?

   _____

3. How many units are you converting?

   _____

4. Use _____.

   $2{,}234 \underline{\phantom{xx}} 10^{\square} =$

   $2{,}234 \underline{\phantom{xxxxx}} = \underline{\phantom{xxxxx}}$

   $2{,}234 \text{ m} = \underline{\phantom{xxxxx}}$

**Guided Practice**

**Convert each measurement.**

**1** 650 mm = _____ dm

A decimeter is _____ units _____ than a millimeter.

650 _____ 10⁻⁻ = _____

**2** 280,900 cm = _____ dkm

A dekameter is _____ units _____ than a centimeter.

280,900 _____ 10⁻⁻ = _____

**3** 3,467 km = _____ m

A meter is _____ units _____ than a kilometer.

3,467 _____ 10⁻⁻ = _____

**4** 7,431 m = _____ dm

A decimeter is _____ unit _____ than a meter.

7,431 _____ 10⁻⁻ = _____

## Step by Step Practice

**5** Convert. 80,743 dm = _____ km

**Step 1** Are you converting to a smaller unit or a larger unit? _____

**Step 2** Do you multiply or divide? _____

**Step 3** How many units are you converting? _____

**Step 4** Use _____.

80,743 _____ 10⁻⁻ = 80,743 _____ = _____

**Step 5** 80,743 dm = _____

**Convert each measurement.**

**6** 10.75 km = _____ hm

_____ by 10⁻⁻.

10.75 _____ 10⁻⁻ = _____

**7** 405 dm = _____ hm

_____ by 10⁻⁻.

405 _____ 10⁻⁻ = _____

GO ON

**Convert each measurement.**

**8** $14.4 \text{ m} = $ _____ mm

$14.4$ _____ $10^{\boxed{\phantom{0}}} = $ _____

**9** $325 \text{ hm} = $ _____ m

$325$ _____ $10^{\boxed{\phantom{0}}} = $ _____

**10** $6.5 \text{ dm} = $ _____ km

$6.5$ _____ $10^{\boxed{\phantom{0}}} = $ _____

**11** $742 \text{ mm} = $ _____ cm

$742$ _____ $10^{\boxed{\phantom{0}}} = $ _____

---

## Step (by) Step **Problem-Solving Practice**

**Solve.**

**12** **SWIMMING**   Reece swims 50 meters in a swim meet. How many hectometers did she swim?

Hectometers are _____ units _____ than meters.

_____ by $10^{\boxed{\phantom{0}}}$.

The decimal moves _____ place(s) to the _____.

_____

Check off each step.

_____ Understand: I underlined key words.

_____ Plan: To solve the problem, I will _____.

_____ Solve: The answer is _____.

_____ Check: I checked my answer by _____.

---

 **Skills, Concepts, and Problem Solving**

**Convert each measurement.**

**13** $45 \text{ dm} = $ _____ mm

**14** $925 \text{ dm} = $ _____ hm

**Convert each measurement.**

**15** 5,487 m = _____ cm

**16** 36,725 mm = _____ dkm

**17** 79 km = _____ hm

**18** 489 dkm = _____ dm

**19** 631 mm = _____ m

**20** 568,734 dm = _____ km

**Solve.**

**21** **RUNNING** Emma ran 11 laps around the track during practice one day. How many kilometers did Emma run during practice?

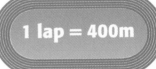
1 lap = 400m

_____

**22** **SCHOOL SUPPLIES** Ethan has new pencils that are each 140 mm long. How long is each pencil in centimeters?

_____

**23** **DRIVING** Garrett drove 62 kilometers to see his grandmother. How many dekameters did he drive?

_____

**Vocabulary Check** Write the vocabulary word that completes each sentence.

**24** The _____ is the base unit for measuring length in the metric system.

**25** The distance between two points is called the _____.

**26** The measuring system based on powers of 10 is the _____.

**27** **Reflect** Write in your own words how to convert between metric units. Do you think it is easier to convert in the metric system or the customary system? Explain your answer.

_____

_____

_____

_____

STOP

**Convert each measurement.**

1   80 oz = _____ lb

2   2 T = _____ lb

3   1.5 lbs = _____ oz

4   1T = _____ oz

5   5,000 cm = _____ m

6   2 km = _____ dkm

7   10 m = _____ cm

8   150 mm = _____ hm

9   200 dkm = _____ cm

10   2,500 mm = _____ m

**Solve.**

11   **ANIMALS**   In a report on giraffes, Sumintra wrote that one animal weighed 1.5. She didn't write down the units. What unit would be appropriate for the weight of a giraffe? How many pounds does the giraffe weigh?

_____

12   **DISTANCE**   Aisha lives 2.5 km from her friend Jason. When she walks to his house, how many meters does she have to walk?

_____

13   **MODEL CARS**   Jacob and Carlos had a contest to see how far their model cars would go on one wind up. Jacob's car went 1.6 m. Carlos' car went 145 cm. Whose car went farther? By how much?

_____

_____

# Capacity in the Metric System

## KEY Concept

The base unit of **capacity** in the **metric system** is the **liter**.

| Unit | Symbol | Number of Liters |
|------|--------|------------------|
| milliliter | mL | 0.001 L |
| centiliter | cL | 0.01 L |
| deciliter | dL | 0.1 L |
| liter | L | 1 L |
| dekaliter | dkL | 10 L |
| hectoliter | hL | 100 L |
| kiloliter | kL | 1,000 L |

To convert between units, multiply or divide by powers of ten. When converting to smaller units, **multiply**. When converting to larger units, **divide**.

A milliliter is one unit smaller than a centiliter.

**3.0 cL = __ mL**

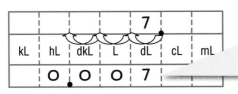

1 cL = 10 cL
Multiplying by 10 moves the decimal point 1 place to the right.

**3 • 10 = 30 mL**

A hectoliter is three units larger than a deciliter.

**7.0 dL = __ hL**

| kL | hL | dkL | L | dL | cL | mL |
|----|----|-----|---|----|----|----|
| | | | | 7 | | |
| | O | O | O | 7 | | |

1 hL = 1,000 dL
Dividing by 1,000 moves the decimal point 3 places to the left.

**7 ÷ 10³ = 7 ÷ 1,000 = 0.007 hL**

The most commonly used units are liter and milliliter. The steps for converting between units of capacity are the same as the steps for converting between units of length.

## VOCABULARY

**capacity**
the amount of dry or liquid material a container can hold

**liter**
a base metric unit for measuring capacity

**metric system**
a measurement system that includes units such as meter, gram, and liter

## Example 1

**Convert. 3.62 kL = ____ L**

1. Are you converting to a smaller unit or a larger unit? smaller

2. Do you multiply or divide? multiply

3. How many units are you converting? three

4. Use $10^3$.

   $3.62 \cdot 10^3 = 3.62 \cdot 1{,}000$

   $= 3{,}620$

   $3.62 \text{ kL} = 3{,}620 \text{ L}$

## YOUR TURN!

**Convert. 61 L = ____ mL**

1. Are you converting to a smaller unit or a larger unit? _____

2. Do you multiply or divide? _____

3. How many units are you converting?

   _____

4. Use _____.

   $61 \, \underline{\quad} \, 10^{\square} = 61 \, \underline{\hspace{2cm}}$

   $= \underline{\hspace{2cm}}$

   $61 \text{ L} = \underline{\hspace{2.5cm}}$

## Example 2

**Convert. 5,813 cL = ____ kL**

1. Are you converting to a smaller unit or a larger unit? larger

2. Do you multiply or divide? divide

3. How many units are you converting?

   five

4. Use $10^5$.

   $5{,}813 \div 10^5 = 5{,}813 \div 100{,}000$

   $= 0.05813$

   $5{,}813 \text{ cL} = 0.05813 \text{ kL}$

## YOUR TURN!

**Convert. 360 mL = ____ dkL**

1. Are you converting to a smaller unit or a larger unit? _____

2. Do you multiply or divide? _____

3. How many units are you converting?

   _____

4. Use _____.

   $360 \, \underline{\quad} \, 10^{\square} = 360 \, \underline{\hspace{2cm}}$

   $= \underline{\hspace{2cm}}$

   $360 \text{ mL} = \underline{\hspace{2.5cm}}$

## ▶ Guided Practice

**Convert each measurement.**

1  437 hL = ____ cL

A centiliter is _____ unit(s)

_____ than a hectoliter.

$437 \, \underline{\quad} \, 10^{\square} = \underline{\hspace{3cm}}$

2  175 dkL = ____ L

A liter is _____ unit(s)

_____ than a dekaliter.

$175 \, \underline{\quad} \, 10^{\square} = \underline{\hspace{3cm}}$

3  Convert.  7,450 cL = _____ hL

**Step 1**  Are you converting to a smaller unit or a larger unit? _____

**Step 2**  Do you multiply or divide? _____

**Step 3**  How many units are you converting? _____

**Step 4**  Use _____. _____ $10^{\boxed{\phantom{x}}}$ = 7,450 _____ = _____

**Step 5**  7,450 cL = _____

**Convert each measurement.**

4  384 mL = _____ dL

_____ by $10^{\boxed{\phantom{x}}}$.

384 _____ $10^{\boxed{\phantom{x}}}$ = _____

5  24,500 dL = _____ kL

_____ by $10^{\boxed{\phantom{x}}}$.

24,500 _____ $10^{\boxed{\phantom{x}}}$ = _____

Step by Step **Problem-Solving Practice**

**Solve.**

6  SOCCER  Camille's soccer team drinks 18 liters of a sports drink during a game. How many milliliters of sports drink does the team drink during a game?

Milliliters are _____ unit(s) _____ than liters.

_____ by $10^{\boxed{\phantom{x}}}$.

_____

The decimal moves _____ place(s) to the _____.

Check off each step.

_____ Understand: I underlined key words.

_____ Plan: To solve the problem, I will _____.

_____ Solve: The answer is _____.

_____ Check: I checked my answer by _____.

GO ON

# Skills, Concepts, and Problem Solving

**Convert each measurement.**

**7** 175 kL = _____ dkL

**8** 321 L = _____ mL

**9** 9.62 cL = _____ hL

**10** 86 hL = _____ L

**11** 5,000 kL = _____ L

**12** 4 mL = _____ cL

**13** 76,500 dkL = _____ cL

**14** 6 hL = _____ mL

**Solve.**

C05-003P-890859.psd

**15** **TRUCKING**   Brock is a truck driver. He used 2,500 liters of gas on his last trip. How many kiloliters of gas did he use?

_____

**16** **MEDICINE**   One dose of a children's medicine is 5 mL. How many centiliters are in one dose?

_____

**17** **SWIMMING**   A school swimming pool holds 375,000 liters of water. How many millimeters of water does it hold?

_____

**Vocabulary Check**   **Write the vocabulary word that completes each sentence.**

**18** A(n) _____ is the base metric unit for measuring capacity.

**19** _____ is the amount of dry or liquid material that a container can hold.

**20** The measuring system that includes milliliters, liters, and kiloliters

is the _____.

**21** **Reflect**   Compare and contrast the units of liter and meter.

_____

_____

_____

STOP

# Mass in the Metric System

## KEY Concept

Mass in the **metric system** is measured using the following units. The base unit of mass in the metric system is the **gram**.

| Unit | Symbol | Number of Meters |
|------|--------|------------------|
| milligram | mg | 0.001 g |
| centigram | cg | 0.01 g |
| decigram | dg | 0.1 g |
| gram | g | 1 g |
| dekagram | dkg | 10 g |
| hectogram | hg | 100 g |
| kilogram | kg | 1,000 g |

To convert between units, multiply or divide by powers of ten. When converting to smaller units, multiply. When converting to larger units, **divide**.

A gram is one unit smaller than a dekagram.

**8.0 dkg = __ g**

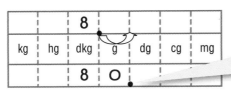

1 dkg = 10 g
Multiplying by 10 moves the decimal point 1 place to the right.

8 • 10 = 80 g

A decigram is two units larger than a milligram.

**2.0 mg = __ dg**

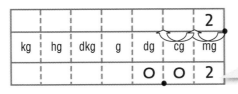

1 dg = 100 mg
Dividing by 100 moves the decimal point 2 places to the left.

$2 \div 10^2 = 2 \div 100 = 0.02$ dg

The most commonly used units for mass are the milligram, gram, and kilogram.

### VOCABULARY

**gram**
a base metric unit for measuring mass

**mass**
the amount of matter in an object

**metric system**
a measurement system that includes units such as meter, gram, and liter

GO ON

## Example 1

**Convert.  1.7 kg = ____ cg**

1. Are you converting to a smaller unit or a larger unit? smaller

2. Do you multiply or divide? multiply

3. How many units are you converting?

   five

4. Use $10^5$.

   $1.7 \cdot 10^5 =$

   $1.7 \cdot 100{,}000 = 170{,}000$

   $1.7 \text{ kg} = 170{,}000 \text{ cg}$

**Convert.  5,662 g = ____ mg**

1. Are you converting to a smaller unit or a larger unit? _____

2. Do you multiply or divide? _____

3. How many units are you converting?

   _____

4. Use _____.

   $5{,}662 \underline{\quad} 10^{\square} =$

   $5{,}662 \underline{\quad\quad} = \underline{\quad\quad}$

   $5{,}662 \text{ g} = \underline{\quad\quad}$

## Example 2

**Convert.  483 cg = ____ dkg**

1. Are you converting to a smaller unit or a larger unit? larger

2. Do you multiply or divide? divide

3. How many units are you converting?

   three

4. Use $10^3$.

   $483 \div 10^3 =$

   $483 \div 100 = 0.483$

   $483 \text{ cg} = 0.483 \text{ dkg}$

**Convert.  3,601.4 mg = ____ g**

1. Are you converting to a smaller unit or a larger unit? _____

2. Do you multiply or divide? _____

3. How many units are you converting?

   _____

4. Use _____.

   $3{,}601.4 \underline{\quad} 10^{\square} =$

   $3{,}601.4 \underline{\quad\quad} = \underline{\quad\quad}$

   $3{,}6014 \text{ mg} = \underline{\quad\quad}$

## ▶ Guided Practice

**Convert each measurement.**

**1**  4,324 cg = ____ g

A gram is _____ units _____ than a centigram.

$4{,}324 \underline{\quad} 10^{\square} = \underline{\quad\quad}$

**2**  6.8 dkg = ____ hg

A hectogram is _____ unit _____ than a dekagram.

$6.8 \underline{\quad} 10^{\square} = \underline{\quad\quad}$

## Step by Step Practice

**3** Convert.   345,000 g = _____ kg

**Step 1**   Are you converting to a smaller unit or a larger unit? _____

**Step 2**   Do you multiply or divide? _____

**Step 3**   How many units are you converting? _____

**Step 4**   Use _____.

345,000 ____ 10$^{\boxed{\phantom{0}}}$ = 345,000 _____ = _____

**Step 5**   345,000 g = _____

**Convert each measurement.**

**4**   640 cg = _____ dkg

_____ by 10$^{\boxed{\phantom{0}}}$.

640 ____ 10$^{\boxed{\phantom{0}}}$ = _____

**5**   3.4 dkg = _____ dg

_____ by 10$^{\boxed{\phantom{0}}}$.

3.4 ____ 10$^{\boxed{\phantom{0}}}$ = _____

## Step by Step Problem-Solving Practice

**Solve.**

**6**   **HEALTH**   A bottle of vitamins has 250 tablets. Each tablet has 200 mg of Vitamin C in it. How many grams of Vitamin C are in one bottle?

Find the total number of milligrams in the bottle.

200 mg • _____ = _____

Convert milligrams to grams. _____ ÷ _____ = _____

Check off each step.

_____ Understand: I underlined key words.

_____ Plan: To solve the problem, I will _____.

_____ Solve: The answer is _____.

_____ Check: I checked my answer by _____.

**GO ON**

# Skills, Concepts, and Problem Solving

**Convert each measurement.**

**7** 36 dkg = _____ g

**8** 2.5 kg = _____ mg

**9** 431 g = _____ cg

**10** 5,320,000 mg = _____ dkg

**11** 600 cg = _____ g

**12** 4,945 dkg = _____ kg

**13** 689 g = _____ mg

**14** 9.25 hg = _____ kg

**Solve.**

**15** **MASS**  Peter's mass is 20.4 kilograms. What is Peter's mass in grams?

_____

**16** **ANIMALS**  The following table lists the mass of three pets. Which pet has the least mass?

_____

| Pet | Mass |
|-----|------|
| Ferret | 20 hectograms |
| Guinea pig | 700 grams |
| Cat | 250 dekagram |

**17** **FOOD**  Jackson buys a bag of pretzels that has a mass of 680 grams. What is the mass of the bag in centigrams?

_____

**Vocabulary Check**  **Write the vocabulary word that completes each sentence.**

**18** The amount of matter in an object is the _____ of the object.

**19** Gram, milligram, and hectogram are all measures of mass in the

_____.

**20** **Reflect**  How can a measure of mass in the metric system give a more precise measurement than a measure of weight in the customary system?

_____

_____

_____

_____

**STOP**

# Perimeter and Area

## KEY Concept

The **perimeter** is the sum of lengths of the sides of a **polygon**. Since opposite sides of a rectangle are equal in length, the formula $P = 2\ell + 2w$ can also be used to find the perimeter of a rectangle.

$P = 7 + 7 + 6 + 6$
   $= 26$ units

or

$P = 2\ell + 2w$
   $= 2(7) + 2(6)$
   $= 26$ units

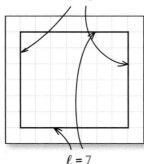

w = 6
ℓ = 7

To find the **area** of a polygon, you can count the number of square units on the inside. The rectangle above contains 42 square units.

Some polygons have special formulas to find the area.

| Object | Area | |
|---|---|---|
| Triangle | $A = \dfrac{1}{2}\,bh$ | (triangle with height $h$ and base $b$) |
| Rectangle | $A = \ell w$ | (rectangle with width $w$ and length $\ell$) |
| Parallelogram | $A = bh$ | (parallelogram with height $h$ and base $b$) |
| Trapezoid | $A = \dfrac{1}{2}\,h(b_1 + b_2)$ | (trapezoid with bases $b_1$, $b_2$ and height $h$) |

Copyright © Glencoe/McGraw-Hill, a division of The McGraw-Hill Companies, Inc.

When using the formulas, first substitute for any variable. Then follow the order of operations to simplify the expression.

GO ON

## VOCABULARY

**area**
the number of square units needed to cover a surface

**perimeter**
the sum of the lengths of the sides of a polygon

**polygon**
a closed plane figure with straight sides

## Example 1

**Find the perimeter.**

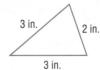

1. Add the length of each side.

$3 + 3 + 2$

2. $P = 8$ in.

## YOUR TURN!

**Find the perimeter.**

1. Add the length of each side.

_____

2. $P = $ _____

## Example 2

**Find the area.**

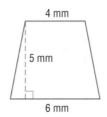

1  Identify the shape.

trapezoid

2. Write the formula for the area.

$A = \frac{1}{2} h(b_1 + b_2)$

3. Name the value for each variable.

$h = 5, b_1 = 4, b_2 = 6$

4. Substitute the values into the formula.

$A = \frac{1}{2} (5)(4 + 6)$

5. Simplify.   $A = \frac{1}{2} \cdot 5(10) = 25$ mm²

## YOUR TURN!

**Find the area.**

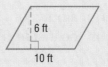

1. Identify the shape.

_____

2. Write the formula for the area.

_____

3. Name the value for each variable.

_____

4. Substitute the values into the formula.

$A = $ _____(_____)

5. Simplify. $A = $ _____(_____) = _____

---

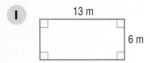 **Guided Practice**

**Find the perimeter of each figure.**

**1**
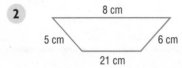

$P = $ ____ + ____ + ____ + ____ = ____

**2**

$P = $ _____ = _____

**3** Find the area of the figure.

**Step 1** Identify the shape. _____

**Step 2** Write the formula for the area. _____

**Step 3** Name the value for each variable. _____

**Step 4** Substitute the values into the formula. _____

**Step 5** Simplify.        $A = $ ___(___) = _____

**Find the area of each figure.**

**4**

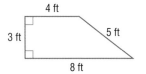

The figure is a _____.

$A = \frac{1}{2} h(b_1 + b_2)$

$A = $ _____

**5**

The figure is a _____.

$A = $ ___

$A = $ _____

**Solve.**

**6** ART PROJECT   Brianne has to make a collage for art class on a poster board that has dimensions 16 inches by 22 inches. Because Brianne has to use the whole board, what is the total area she needs to cover with her collage?

The board is a _____.

$A = $ _____

Check off each step.

_____ Understand: I underlined key words.

_____ Plan: To solve the problem, I will _____.

_____ Solve: The answer is _____.

_____ Check: I checked my answer by _____.

GO ON

# ▶ Skills, Concepts, and Problem Solving

**Find the perimeter and the area of each figure.**

**7**
9 ft
11 ft

_____

**8**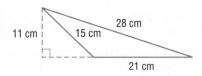
11 cm    15 cm    28 cm
21 cm

_____

**9**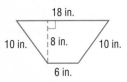
18 in.
10 in.    8 in.    10 in.
6 in.

_____

**10**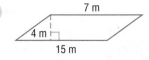
7 m
4 m
15 m

_____

## Solve.

**11**  **TENNIS**   A rectangular tennis court is 78 ft long and 36 ft wide. What is the area of the tennis court?

_____

**12**  **FLAGS**   Sonja makes nylon flags of different shapes and sizes. One of her customers wants a flag in the shape shown to the right. How much nylon will Sonja need for this flag?

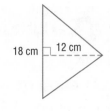
18 cm    12 cm

**Vocabulary Check**   **Write the vocabulary word that completes each sentence.**

**13**  A closed plane figure is called a(n) _____.

**14**  The _____ is the sum of the lengths of the sides of a polygon.

**15**  _____ is the number of square units needed to cover a surface.

**16**  **Reflect**   Kirk knows the area of a rectangle and the length of one side of the rectangle. Does he have enough information to find the perimeter? Explain.

_____

_____

_____

**STOP**

## Convert each measurement.

**1**  300 cL = _____ L

**2**  6,000 mg = _____ g

**3**  4 kL = _____ liters

**4**  16 hL = _____ cL

**5**  12 g = _____ kg

**6**  11 hg = _____ g

**7**  150 kg = _____ dg

**8**  19 dL = _____ hL

## Find the perimeter and the area of each figure.

**9**
15 m
25 m

_____

**10**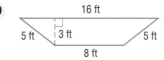
16 ft
5 ft   3 ft   5 ft
8 ft

_____

**11**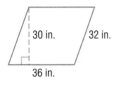
30 in.   32 in.
36 in.

_____

**12**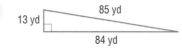
85 yd
13 yd
84 yd

_____

## Solve.

**13**  **GROCERIES**   Mehlia bought 6,804 g of fruit at the grocery. How many kilograms of fruit did she buy?

_____

**14**  **HEALTH**   Doctors recommend that adults drink about 1,900 mL of water every day. About how many liters of water should an adult drink per day?

_____

**15**  **SCIENCE**   In an experiment, Reina measured 145 mL of hydrochloric acid. How many liters did she have?

_____

**Convert each measurement.**

**1** 1 km = _____ hm

1 hm = _____ dkm

1 dkm = _____ m

1 m = _____ dm

1 dm = _____ cm

1 cm = _____ mm

**2** 1 kg = 10 _____

1 hg = 10 _____

1 dkg = 10 _____

1 g = 10 _____

1 dg = 10 _____

1 cg = 10 _____

**3** 1 kL = 10 _____

1 hL = _____ dkL

1 _____ = 10 _____

1 L = 10 _____

1 dL = _____ cL

1 _____ = 10 mL

**4** 15 dkL = _____ mL

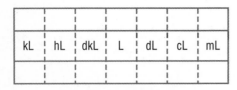

| kL | hL | dkL | L | dL | cL | mL |
|----|----|-----|---|----|----|----|
| | | | | | | |

**5** 4,120 g = _____ kg

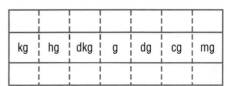

| kg | hg | dkg | g | dg | cg | mg |
|----|----|-----|---|----|----|----|
| | | | | | | |

**6** 192 in. = _____ ft

**7** 1.7 km = _____ cm

**8** 1,200 c = _____ gal

**9** 4 T = _____ oz

**10** 16 c = _____ fl oz

**11** 5 mi = _____ ft

**12** 72 oz = _____ lb

**13** 6.7 g = _____ mg

**14** 800 L = _____ cL

**15** 328 mm = _____ m

**Find the perimeter and area of each figure.**

**16**
52 ft  52 ft
10 ft
48 ft

_____

**17**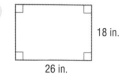
18 in.
26 in.

_____

**18**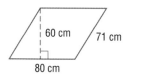
60 cm  71 cm
80 cm

_____

**19**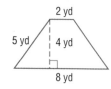
2 yd
5 yd  4 yd
8 yd

_____

**Solve.**

**20 TRAVEL** Karina's family drove from Niagara Falls to Toronto, Canada. The distance was 133.23 km. How far was it from Niagra Falls to Toronto in centimeters?

_____

**21 SWIMMING** There are 8,358,000 liters of water in Jerald's pool. How many kiloliters of water are in the pool?

_____

**22 LANDSCAPING** Tyrell created a rectangular flower bed that was 12 ft long and 8 ft wide. One bag of mulch covers 3 square feet. How many bags of mulch did Tyrell need for his flower bed?

_____

**Correct the mistake.**

**23** Melina found the area of the figure shown to be 32.9 m². Determine what mistake Melina made and find the correct area.

2.8 m  3.5 m
9.4 m

_____

_____

# Probability and Statistics

## Statistics are often used to measure data and data trends.

The birth rate in the United States in 2007 was 13.8 births for every 1,000 women. The birth rate varies from country to country and from year to year.

STEP **2** **Preview**   Get ready for Chapter 6. Review these skills and compare them with what you will learn in this chapter.

| What You Know | What You Will Learn |
|---|---|
| You can find the value of the following expression. $$(12 + 15 + 13 + 16) \div 4$$ $$= (56) \div 4$$ $$= 14$$ **TRY IT!** **1** $(11 + 6 + 32 + 9) \div 4 =$ _____ | *Lesson 6-1* The **mean** of a set of data is the sum of the data divided by the number of pieces of data. Data set: 4, 6, 9, 13 Mean: $\dfrac{4 + 6 + 9 + 13}{4} = \dfrac{32}{4} = 8$ |

You can choose a slice of pizza and a drink from the menu shown.

| Pizza | Drink |
|---|---|
| cheese | lemonade |
| pepperoni | fruit punch |
| vegetable | |

**Example:** Describe two different choices a person could make.

Choice 1: cheese pizza and lemonade
Choice 2: pepperoni pizza and lemonade

**TRY IT!**

**2** Describe three more choices.

_____

_____

_____

*Lesson 6-2*

A **tree diagram** is an organized way to list all the possible outcomes when given multiple events.

Find the number of possible outcomes for choosing a slice of pizza and a drink from the menu at the left.

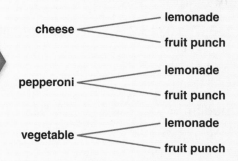

There are 6 combinations of 1 slice of pizza and 1 drink.

# Mode, Median, and Mean

## KEY Concept

The **mode**, **median**, and **mean** are measures of central tendency that describe a set of **data**.

### Mode

The mode is the number or numbers that occur most often. There can be one mode, no mode, or many modes.

4, **22**, 14, **22**, 18          The mode is **22**.

occurs two times

### Median

To find the median, list data in numerical order. The median is the middle number in a set of data that is arranged in numerical order.

4, 22, 14, 22, 18          The median is **18**.

4   14   **18**   22   22

middle number

With an even number of data, the median is the mean of two middle numbers.

4, 6, **7, 10,** 13, 15          The median is **8.5**.

middle numbers

$$\text{median} = \frac{7 + 10}{2} = \frac{17}{2} = 8.5$$

### Mean

To find the mean, first find the sum of all data in the set. Then divide by the number of elements in the set.

4, 22, 14, 22, 18          The mean is **16**.

$$\text{mean} = \frac{4 + 14 + 18 + 22 + 22}{5} = \frac{80}{5} = 16$$

## VOCABULARY

**data**
information collected from a survey or experiment

**mean**
the sum of numbers in a set of data divided by the number of items in the data set

**median**
the middle number in a set of data when the data are arranged in numerical order. If the data set has an even number, the median is the mean of the two middle numbers.

**mode**
the number(s) that appear most often in a set of data

## Example 1

**Find the median and mode.**

$$16, 23, 17, 27, 19, 20, 26, 20$$

1. Find the median.

   List the data in order. Circle the middle number(s).

   $$16, 17, 19, \widehat{20, 20}, 23, 26, 27$$

   $$\frac{20 + 20}{2} = 20$$

2. Find the mode.

   Look for any numbers that are listed more than one time.

   **20 is listed two times.**

3. The median is 20.

   The mode is 20.

**YOUR TURN!**

**Find the median and mode.**

$$54, 60, 58, 63, 68, 71, 74, 60, 68$$

1. Find the median.

   List the data in order. Circle the middle number(s).

   ____, ____, ____, ____, ____, ____, ____,

   ____, ____

2. Find the mode.

   Look for any numbers that are listed more than one time.

   ____ is listed two times.

   ____ is listed two times.

3. The median is ____.

   The modes are ____ and ____.

## Example 2

**Find the mean of the data.**

$$3, 14, 0, 8, 10, 12, 7, 15$$

1. Find the sum of the data.

   $$3 + 14 + 0 + 8 + 11 + 12 + 7 + 15 = 70$$

2. Count the items in the data set.

   **8**

3. Divide the sum by 8.

4. The mean is 8.75.

$$
\begin{array}{r}
8.75 \\
8{\overline{)70.00}} \\
-64\phantom{.00} \\
\hline
60\phantom{.0} \\
-56\phantom{.0} \\
\hline
40 \\
-40 \\
\hline
0
\end{array}
$$

**YOUR TURN!**

**Find the mean of the data.**

$$20, 43, 35, 24, 28$$

1. Find the sum of the data.

   _____

2. Count the item in the data set.

   _____

3. Divide the sum by _____.

4. The mean is _____.

**GO ON**

**Guided Practice**

**Find the median and mode of each data set.**

**1**   15, 8, 13, 22, 8, 13

_____, _____, _____, _____,

_____, _____

median: _____

mode: _____, _____

**2**   40, 78, 12, 34, 90, 61, 78, 36, 42

_____, _____, _____, _____, _____,

_____, _____, _____, _____

median: _____

mode: _____

**Step** (by) **Step Practice**

**3**   Find the mean of the data.

      **14, 8, 20, 24, 8, 15, 26, 14, 8, 19, 18, 12**

**Step 1**   Find the sum of the data.

_____

**Step 2**   Count. There are _____ items in the data set.

**Step 3**   Divide the sum by _____.

**Step 4**   The mean is _____.

**Find the mean of each data set.**

**4**   41, 75, 9, 25, 65, 61, 101, 53, 29

$$\text{mean} = \frac{41 + 75 + 9 + 25 + 65 + 61 + 101 + 53 + 29}{\boxed{\phantom{x}}} = \frac{\boxed{\phantom{x}}}{\boxed{\phantom{x}}} = \underline{\phantom{xxx}}$$

**5**   12, 6, 9, 36, 4, 15, 86

$$\text{mean} = \frac{\boxed{\phantom{xxxxxxxxxxxxxxxxxxxxxxxx}}}{\boxed{\phantom{x}}} = \frac{\boxed{\phantom{x}}}{\boxed{\phantom{x}}} = \underline{\phantom{xxx}}$$

**Find the mean of each data set.**

**6** 87, 93, 27, 15, 27, 38, 7, 12, 45

$$\dfrac{\boxed{\phantom{xxxxxxxxxxxxxxxxxxx}}}{\boxed{\phantom{x}}} = \dfrac{\boxed{\phantom{x}}}{\boxed{\phantom{x}}} = \underline{\phantom{xxxxx}}$$

**7** 19, 30, 84, 26, 30, 71, 105, 84, 57, 66

$$\dfrac{\boxed{\phantom{xxxxxxxxxxxxxxxxxxx}}}{\boxed{\phantom{x}}} = \dfrac{\boxed{\phantom{x}}}{\boxed{\phantom{x}}} = \underline{\phantom{xxxxx}}$$

## Step by Step Problem-Solving Practice

**Solve.**

**8** **BOWLING** Venkat's bowling scores for a 3-day tournament were 204, 178, 155, 219, 196, 204, 188, 183, and 210. Find the mean, median, and mode of his scores.

List the data in order: ____, ____, ____, ____, ____, ____, ____, ____, ____

mean = _____

median: _____    mode: _____    mean: _____

Check off each step.

_____ Understand: I underlined key words.

_____ Plan: To solve the problem, I will _____.

_____ Solve: The answer is _____.

_____ Check: I checked my answer by _____.

## ▶ Skills, Concepts, and Problem Solving

**Find the median, mode, and mean of each data set.**

**9** 90, 44, 41, 50, 56, 49, 41

median: _____

mode: _____

mean: _____

**10** 12, 13, 7, 5, 25, 25, 5, 10, 17, 16, 19

median: _____

mode: _____

mean: _____

GO ON

**Find the median, mode, and mean of each data set.**

**11**  20, 73, 31, 53, 22, 64, 45

median: _____

mode: _____

mean: _____

**12**  55, 33, 48, 6, 9, 55, 27, 56, 8, 80

median: _____

mode: _____

mean: _____

**13**  107, 143, 96, 116, 130, 128

median: _____

mode: _____

mean: _____

**14**  49, 59, 40, 34, 24, 17, 57, 40

median: _____

mode: _____

mean: _____

**Solve.**

**15**  **HOUSING**   Five houses in Aida's neighborhood have sold in the last three months. They sold for $140,000, $175,000, $159,000, $144,000 and $154,000. Aida wants to know the mean selling price and the median selling price of the houses that have sold.

_____

**16**  **TICKET PRICES**   Rashid's favorite band is coming to town. He sees the sign at the right at the ticket office. Each section has the same number of seats. What is the mean ticket price for the concert?

_____

| Ticket Prices | | | |
|---|---|---|---|
| Section A | $95 | Section C | $70 |
| Section B | $75 | Section D | $55 |

**Vocabulary Check**   **Write the vocabulary word that completes each sentence.**

**17**  The _____ is the middle score in a set of data.

**18**  The most frequent score in a set of data is the _____.

**19**  The mean of the set of data is the same as the _____ of the set.

**20**  **Reflect**   If a set of data has one number much greater than the rest of the data, how does that affect the mean and median of the data?

_____

_____

# Count Outcomes

## KEY Concept

There is more than one way to determine the number of possible outcomes when given multiple events.

### Tree Diagrams

A **tree diagram** is an organized way to list the possible **outcomes** when given multiple events.

The possible outcomes of choosing one of three entrées and one of two side dishes are shown in the tree diagram.

| Entrée | Side | Outcomes |
|---|---|---|
| taco | rice | taco, rice |
| | beans | taco, beans |
| burrito | rice | burrito, rice |
| | beans | burrito, beans |
| enchilada | rice | enchilada, rice |
| | beans | enchilada, beans |

### The Fundamental Counting Principle

If event $A$ can happen $a$ ways and event $B$ can happen $b$ ways, then the number of ways event $A$ followed by event $B$ can occur is $a \cdot b$.

The number of ways of choosing one of three entrées and one of two side dishes is $3 \cdot 2 = 6$.

### Factorials

To find the number of ways of arranging items in any order, use a **factorial**.

$$n! = n(n - 1)(n - 2)\ldots(1)$$

The number of ways of arranging 4 books on a shelf in any order is 4! (This is read "four factorial.")

$$4! = 4(4 - 1)(4 - 2)(4 - 3)$$
$$= 4 \cdot 3 \cdot 2 \cdot 1$$
$$= 24$$

### VOCABULARY

**factorial**
the expression $n!$, read $n$ factorial, where $n$ is greater than zero, is the product of all positive integers beginning with $n$ and counting backward to 1

**Fundamental Counting Principle**
if an event $M$ can occur in $m$ ways and is followed by an event $N$ that can occur in $n$ ways, then the event $M$ followed by the event $N$ can occur in $m \times n$ ways

**outcome**
one possible result of a probability event

**tree diagram**
a diagram used to show the total number of possible outcomes

A tree diagram, the Fundamental Counting Principle, or a factorial can be used to calculate the number of outcomes in an experiment.

## Example 1

**Draw a tree diagram to find the number of outfits Pablo can make from a blue, brown, or white shirt and navy, tan, or black pairs of pants.**

1. Make a tree diagram.

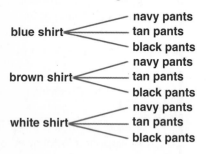

blue shirt — navy pants
— tan pants
— black pants

brown shirt — navy pants
— tan pants
— black pants

white shirt — navy pants
— tan pants
— black pants

2. Count. There are 9 combinations of one pair of pants and one shirt.

## YOUR TURN!

**Draw a tree diagram to find the number of cars Lucy can order from sedan, or luxury models and colors of red, white, black, or silver.**

1. Make a tree diagram.

sedan

luxury

2. Count. There are _____ combinations of one car model and one color.

## Example 2

**Use the Fundamental Counting Principle to find the number of possible outcomes.**

A carnival game has 5 doors. Behind each door are 2 curtains. Find the number of ways a player can choose one door and then one curtain.

> This is event A.

1. There are 5 doors.

> This is event B.

2. There are 2 curtains.

3. Use the Fundamental Counting Principle.
   5 • 2 = 10

4. There are 10 ways to choose one door and then one curtain.

## YOUR TURN!

**Use the Fundamental Counting Principle to find the number of possible outcomes.**

Kenyi has 3 ways to get from home to the grocery store. From the grocery store, she has 6 ways to get to the bank. Find the number of possible ways Kenyi can go from home to the grocery store and then to the bank.

1. There are _____ ways to the grocery store.

2. There are _____ ways to the bank.

3. Use the Fundamental Counting Principle.

   _____ • _____ = _____

4. There are _____ ways Kenyi can go from home to the grocery store, and then the bank.

## Example 3

Jessica, Sally, Yu-Jun, Allie, and George are waiting in line. Use a factorial to find how many different ways they can stand in line.

1. Five people can be arranged in any order.

2. Use 5! to find the number of ways the people can stand in line.

$5 \cdot 4 \cdot 3 \cdot 2 \cdot 1 = 120$

### YOUR TURN!

Use a factorial to find the number of 7-digit passcodes that can be made from the numbers 1, 2, 3, 4, 5, 6, and 7.

1. _____ can be arranged in any order.

2. Use _____ to find number of passcodes.

_____

 **Guided Practice**

**Draw a tree diagram to find the number of possible outcomes for each situation.**

1. Yvonne can choose a new white, beige, or stainless steel kitchen sink system and a one-handled faucet or two-handled faucets.

white

beige

stainless steel

There are _____ combinations of one sink and one faucet.

2. Boyd High School is choosing a design for a school flag. The section labeled 1 can be blue or red. The section labeled 2 can be orange or yellow. The section labeled 3 must be black.

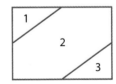

blue

red

There are _____ combinations of colors for the flag.

GO ON

**Use the Fundamental Counting Principle to find the number of possible outcomes.**

3   Zoe and Selena are ordering a pizza from the menu at the right. They decide not to consider olives, banana peppers, or bacon as toppings. How many different topping combinations could they order if they want one meat topping and one vegetable topping?

| Pizza Menu | |
|---|---|
| Meat Toppings | Vegetable Toppings |
| Bacon | Banana Peppers |
| Ham | Green Peppers |
| Pepperoni | Mushrooms |
| Sausage | Olives |
| | Onions |
| | Spinach |

**Step 1**   There are _____ meat choices.

**Step 2**   There are _____ vegetable choices.

**Step 3**   Use the Fundamental Counting Principle.

_____ • _____ = _____

**Step 4**   Zoe and Selena can order _____ topping combinations on the pizza.

**Use the Fundamental Counting Principle to find the number of possible outcomes.**

4   Ramiro has 5 shirts and 3 ties in his closet. Find the number of ways he can choose one shirt and one tie.

_____ • _____ = _____

5   A school is selling spirit shirts in the colors of white, gray, and blue. The styles are short sleeve, long sleeve, and polo shirts. How many different styles of shirts are for sale?

_____ • _____ = _____

6   Marcelle and her family are ordering pasta for dinner. They can choose from garlic bread or french bread. They have choices of spaghetti, lasagna, ziti, or fettuccine. How many different combinations could they make?

_____ • _____ = _____

**Use a factorial to find the number of possible outcomes.**

**7** Using the digits 1, 2, 3, and 4, how many 4-digit pin codes can be made?

4 (4 − _____) (4 − _____) (4 − _____)

4 · _____ · _____ · _____ = _____

## Step by Step Problem-Solving Practice

**Solve.**

**8** SLEDDING    There are 6 children sledding on a toboggan. How many different ways can they sit on the toboggan?

_____ · _____ · _____ · _____ · _____ · _____ = _____

Check off each step.

_____ **Understand: I underlined key words.**

_____ **Plan: To solve the problem, I will** _____.

_____ **Solve: The answer is** _____.

_____ **Check: I checked my answer by** _____.

## Skills, Concepts, and Problem Solving

**Draw a tree diagram to find the number of possible outcomes for each situation.**

**9** 3 books and 2 magazines; choose one of each

**10** a new car with 4 exterior color and 3 interior fabric choices

_____           _____

**GO ON**

**Use the Fundamental Counting Principle to find the number of possible outcomes.**

**11** a dinner if there are 5 choices for an entrée and 4 choices for a side dish

_____

**12** 4 cushions and 4 stools; choose one of each

_____

**Use a factorial to find the number of possible outcomes.**

**13** a 5-letter pass code using only vowels

_____

**14** 7 classes in a seven-period school day

_____

**Solve.**

**15** SCHOOL LUNCH   The sign to the right is posted at the school cafeteria. How many lunch combinations are possible?

_____

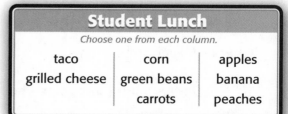

| Student Lunch | | |
| Choose one from each column. | | |
| taco | corn | apples |
| grilled cheese | green beans | banana |
| | carrots | peaches |

**16** STUDENT GOVERNMENT   Four students are running for the offices of president, vice president, secretary, and treasurer of their class. How many different ways can the students fill the offices?

_____

**Vocabulary Check**   **Write the vocabulary word that completes each sentence.**

**17** One possible result of a probability event is a(n) _____.

**18** A _____ is a model used to show the total number of possible outcomes.

**19** [Reflect]   Compare the Fundamental Counting Principle and a tree diagram.

_____

_____

_____

STOP

# Probability

## KEY Concept

To calculate the **probability** of an event, divide the number of favorable **outcomes** by the total possible outcomes. A favorable outcome is a desired event.

$$\text{Probability} = P(\text{event}) = \frac{\text{favorable outcomes}}{\text{possible outcomes}}$$

When a number cube is rolled, there are 6 possible outcomes: 1, 2, 3, 4, 5, 6. The probability of an even number in this event has 3 favorable outcomes.

$$P(\text{even number}) = \frac{\text{favorable outcomes}}{\text{possible outcomes}} = \frac{3}{6} = \frac{1}{2} = 0.5$$

A probability of 0.5 (or $\frac{1}{2}$) means the event is as equally likely to happen as not to happen.

Event is **impossible** to happen. 0

Event is **certain** to happen. 1

$\frac{1}{2}$

0.5

### VOCABULARY

**outcome**
one possible result of a probability event

**probability**
the ratio of the number of favorable equally likely outcomes to the number of possible equally likely outcomes

Probability can be written as a fraction or a decimal.

## Example 1

A bag contains 14 blue marbles and 21 red marbles. If one marble is chosen at random, what is the probability that it will be red?

1. Find the number of favorable outcomes.

   **There are 21 red marbles in the bag.**

2. Find the number of possible outcomes.

   **14 + 21 = 35 marbles**

3. Find the probability.

   $$P(\text{red}) = \frac{\text{favorable outcomes}}{\text{possible outcomes}} = \frac{21}{35}$$

   $$\frac{21}{35} \div \frac{7}{7} = \frac{3}{5} = 0.6$$

## YOUR TURN!

A class has 12 girls and 13 boys. If one student is chosen at random, what is the probability that the student will be a boy?

1. Find the number of favorable outcomes.

   There are _____.

2. Find the number of possible outcomes.

   _____

3. Find the probability.

   $$P(\_\_\_\_) = \frac{\text{favorable outcomes}}{\text{possible outcomes}} = \frac{\square}{\square}$$

   $$\frac{\square}{\square} = \_\_\_\_$$

GO ON

## Example 2

The spinner is spun once. What is the probability that the arrow lands on blue?

1. Find the number of favorable outcomes.
   There is **1 section** colored blue.

2. Find the number of possible outcomes.
   There are **4 sections** on the spinner.

3. Find the probability.

   $P(\text{blue}) = \dfrac{\text{favorable outcomes}}{\text{possible outcomes}}$

   $= \dfrac{1}{4} = 0.25$

### YOUR TURN!

A bridge is built in four equal sections. If a car travels across the bridge, what is the probability at any moment the car is on a steel section?

| concrete | steel | steel | concrete |
|----------|-------|-------|----------|

1. Find the number of favorable outcomes.
   There are _____ sections.

2. Find the number of possible outcomes.
   There are _____.

3. Find the probability.

   $P(\underline{\hspace{1cm}}) = \dfrac{\text{favorable outcomes}}{\text{possible outcomes}}$

   $= \dfrac{\square}{\square} \div \dfrac{\square}{\square} = \dfrac{\square}{\square} = \underline{\hspace{1cm}}$

## ▶ Guided Practice

**Find the probability.**

1. What is the probability of rolling a number greater than 3 on a standard number cube?

   There are _____ numbers greater than 3 on a number cube. There are _____ numbers on a number cube.

   $P(n > 3) = \dfrac{\text{favorable outcomes}}{\text{possible outcomes}} = \underline{\hspace{2cm}}$

## Step by Step Practice

2. Discs numbered from −6 to 3 are in a bag. If one disc is chosen at random, what is the probability of selecting a disc with a negative number?

   **Step 1** _____ outcomes: There are _____ discs with negative numbers.

   **Step 2** _____ outcomes: There are _____ discs in the bag.

   **Step 3** $P(\text{negative number}) = \dfrac{\text{favorable outcomes}}{\text{possible outcomes}} = \underline{\hspace{1cm}} = \underline{\hspace{1cm}}$

**3** What is the probability of spinning an even number on the spinner?

There are _____ even numbers: _____ and _____.

There are _____ numbers on the spinner.

$P(\text{even}) = \dfrac{\text{favorable outcomes}}{\text{possible outcomes}} =$ _____

**A bag contains 10 red marbles, 26 green marbles and 4 orange marbles. Find each probability.**

**4** choosing a red marble

$P(\rule{3em}{0.4pt}) =$ _____

**5** choosing a green marble

_____

**6** choosing an orange marble

_____

## Step (by) Step *Problem-Solving Practice*

**Solve.**

**7** QUIZZES   Mr. Simon gives one pop quiz each week. The students in Mr. Simon's history class knows that he is equally likely to give a pop quiz on any day of the week. What is the probability that Mr. Simon will give a pop quiz on Friday?

favorable outcome: _____

possible outcomes: _____

$P(\rule{6em}{0.4pt}) =$ _____

Check off each step.

_____ Understand: I underlined key words.

_____ Plan: To solve the problem, I will _____.

_____ Solve: The answer is _____.

_____ Check: I checked my answer by _____.

GO ON

**Skills, Concepts, and Problem Solving**

**Find each probability.**

**8** selecting a month that does not begin with the letter "J"

_____

**9** choosing a girl at random, from a class with 7 boys and 13 girls

_____

**10** drawing a red marble from a bag with 18 black marbles, 9 red and 3 yellow marbles

_____

**11** drawing a yellow marble from a bag with 17 black marbles, 12 red and 21 yellow marbles

_____

**Solve.**

**12** **SOCCER**   There are 9 girls and 3 boys on a soccer team. If a captain for a game is chosen at random, what is the probability the captain is a boy?

_____

**13** **GAMES**   To move a player on a board game, you have to spin the spinner. If Gus needs a number greater than 5 to win, what is the probability he will spin a number large enough to win?

_____

**Vocabulary Check**   **Write the vocabulary word that completes each sentence.**

**14** An outcome that is desired is a(n) _____ for an experiment.

**15** The probability of an event is a number between _____.

**16** The probability of an event is the ratio of favorable outcomes

to the _____.

**17** **Reflect**   What does a probability of 1.0 mean? What does a probability of 0 mean?

_____

_____

**Find the median, mode, and mean of each data set.**

**1** 19, 16, 16, 18, 17, 16, 17

median: _____

mode: _____

mean: _____

**2** 48, 51, 30, 41, 51, 50, 41, 49, 53

median: _____

mode: _____

mean: _____

**3** 4, 2, 10, 6, 3

median: _____

mode: _____

mean: _____

**4** 83, 68, 90, 95, 74, 80, 72, 88

median: _____

mode: _____

mean: _____

**Draw a tree diagram to find the number of possible outcomes for each situation.**

**5** Two choices of taco shells: crunchy or soft; Two choices of fillings: chicken or beef

_____ outcomes

**6** Two types of chain: silver or gold Three pendants: ruby, sapphire, or diamond

_____ outcomes

**A box of fruit chews contains 12 apple-flavored, 28 pear-flavored, 36 banana-flavored, and 4 lemon-flavored chews. Find each probability, assuming the chews are randomly selected.**

**7** $P$(apple) _____

**8** $P$(banana) _____

**Solve.**

**9** **BAKING** A bakery offers three choices for birthday cakes: white, yellow, or chocolate. They also offer five themes: balloons, clowns, fireworks, dolls, and bears. How many possible types of cakes are available?

_____

# Chapter 6

# Chapter Test

**Find the median, mode, and mean for each data set.**

**1**  50, 48, 39, 43, 51, 45, 42, 47

median: _____

mode:  _____

mean:  _____

**2**  11, 18, 23, 11, 18, 20, 14, 17, 21

median: _____

mode:  _____

mean:  _____

**3**  70, 71, 61, 67, 68, 63, 72, 73, 63, 65

median: _____

mode:  _____

mean:  _____

**4**  150, 160, 162, 151, 156, 168, 145

median: _____

mode:  _____

mean:  _____

**Draw a tree diagram to find number of possible outcomes for each situation.**

**5**  three types of pants: shorts, Capri, dress
two colors: black, khaki

**6**  two pictures: individual, group
four frame finishes: black, brown,
silver, gold

_____ outcomes

_____ outcomes

**Use the Fundamental Counting Principle to find the number of possible outcomes.**

**7**  3 types of ice cream and 4 types
of toppings; choose one of each

**8**  5 models of cars, 10 colors from which
to choose

_____

_____

**9**  4 doll dresses, 3 hats; 3 types
of shoes; choose one of each

**10**  4 types of pens in 5 colors

_____

_____

## Find each probability.

**11** randomly selecting a white pair of socks from a drawer with 9 pairs of white socks and 11 pairs of black socks

_____

**12** randomly selecting a yellow chip from a pile of 12 yellow chips, 10 blue chips, and 18 red chips

_____

**13** randomly selecting a red marble from a bag with 7 yellow marbles, 3 red marbles, and 40 blue marbles

_____

**14** randomly selecting a day of the week that contains the letters a, d, and y

_____

## Solve.

**15** **SHOPPING**   A clothing store has 3 styles of pants in 4 colors. The pants come with a choice of 2 different designs or no design on them. How many types of pants are available?

_____

**16** **VIDEO GAMES**   Jonathan's video game scores for 5 games were 561, 527, 589, 421, 602. What was his mean score for these 5 games?

_____

**17** **CONTESTS**   In order to win the grand prize, Maria must spin the prize wheel once and land on "vacation." What is the probability that she will win the grand prize?

_____

## Correct the mistake.

**18** **REAL ESTATE**   In Devon's community, several houses just sold. He recorded the selling price of each house in the following table. Based on his table, he says that the median house price is $234,400. Is he correct?

_____

_____

_____

| Address | Price |
|---|---|
| 211 Apple St. | $189,000 |
| 254 Apple St. | $145,900 |
| 65 Orange Ave. | $215,500 |
| 5518 Lemon Way | $234,400 |
| 3522 Lemon Way | $175,800 |
| 1639 Orchard Dr. | $205,500 |
| 1825 Orchard Dr. | $199,500 |

number properties, 4–8

numbers
compare and order real, 63–66
rational and irrational, 38

numerator, 43

operation, 25

order real numbers, 63–66

ordered pair, 108

order of operations, 25–28, 77, 93

outcome(s)
count, 191–196
defined, 191, 197

parallelogram, area of, 177

parentheses, 25

percent(s), 49, 53
and decimals, 49–52
and fractions, 53–57

perfect square, 59

perimeter, 177–180
defined, 177
in real-world applications, 136

polygon, 177

powers of ten, 163

Probability, 191–200

problem-solving. See Step-by-Step Problem Solving

product, 15

Progress Check, 14, 24, 33, 48, 58, 67, 82, 92, 103, 120, 132, 143, 158, 168, 181, 201

properties of numbers, 4–8

proportion, 148, 153, 159
conversions in, 153–157

Quizzes. See Progress Check

quotient, 15

R

radical sign, 59

range, 108

ratio, 49

rational numbers, 38–42, 63

real numbers, 37, 38, 63
compare and order, 63–66

Real-World Applications
accessibility, 145
advertising, 24
allowance, 12
angles, 81
animals, 168, 176
area, 18, 23, 28, 32, 33, 35, 42, 61, 62, 76, 91, 102
art project, 179
baking, 201
banking, 13, 35
baseball, 42, 114, 152, 195
basketball, 46, 52
bowling, 101, 189
car rental, 86
cars, 23
cement, 114
clothes, 69
collecting, 82
comic books, 24
computers, 62
contests, 203
cooking, 28, 48
decorating, 67
dieting, 14
distance, 168
driving, 167
elections, 143
exercise, 57, 58
fencing, 151
financial literacy, 12, 13, 18, 35, 75, 82, 91, 118, 131, 142, 162, 190, 203
flags, 180
food, 176
football, 13
fundraising, 82
games, 48, 200
gardening, 69
geometry, 105
grades, 52, 82
graphing, 126
grocery, 157, 181

health, 161, 175, 181
hiking, 33
home improvement, 92, 145
housing, 190
hygiene, 58
integers, 98
juice, 52
knitting, 152
landscaping, 112, 119, 183
law enforcement, 158
logic, 28
mail, 14
manufacturing, 105
mass, 176
measurement, 66, 67, 152
media, 62
medicine, 103, 172
model cars, 168
money, 142
number sense, 85, 86, 135
painting, 32
party planning, 81, 103, 158
payday, 18
perimeter, 136
pets, 120, 143
pizza, 56
postcards, 22
printing, 156
quality control, 65
quizzes, 199
real estate, 203
rental car, 102
restaurant, 57, 162
roads, 119
running, 8, 167
sales, 91, 131
savings, 92, 132
school, 41, 51
school lunch, 157, 196
school play, 23
school supplies, 18, 167
science, 181
sewing, 158
shopping, 75, 162, 203
soccer, 171, 200
sports, 105
student government, 196
survey, 47
swimming, 166, 172, 183
temperature, 31, 98, 105
tennis, 180
testing, 47, 76
tests, 136
test scores, 7, 97